H. NORMAN SCHWARZKOPF —A REAL AMERICAN HERO

"A warrior to be proud of." —*Time* magazine

"A warrior with a soul." —*Newsweek*

"It is [his] rare blend—martial mastery and human sensitivity—that draws raves from friends and foot soldiers." —*People*

"I'd like to think of myself as a man of conscience."
—General H. Norman Schwarzkopf

M. E. Morris is a 30-year veteran of the U.S. Navy, a pilot, graduate of the Naval War College, and a retired Navy captain. The holder of a Masters degree in International Relations, Captain Morris served as a Designated Subspecialist in International Affairs for the Navy. He is the author of several nonfiction military books and five novels.

H. NORMAN SCHWARZKOPF

ROAD TO TRIUMPH

CAPT. M.E. MORRIS, U.S.N. (RET.)

Illustrations by Harry Trumbore

SCHWARZKOPF: ROAD TO TRIUMPH

Cover photograph by Neil Greentree, Sepa Press.
Maps and illustrations drawn by Harry Trumbore.

ISBN: 0-312-92753-3

Printed in the United States of America

St. Martin's Paperbacks edition/June 1991

10 9 8 7 6 5 4 3 2 1

Respectfully, for The Bear and
the men and women of
DESERT STORM

ACKNOWLEDGMENTS

This work reflects the professional input of a number of people. First, it was the extraordinary confidence of my agent, Jane Dystel (Acton & Dystel, Inc.) that led her to recommend me for the project. Tom Dunne, the splendid executive editor who put this all together, placed an enormous amount of trust in me by accepting a more or less sight-unseen author. His guidance and encouragement carried us over the rough spots with amazing grace. Pete Wolverton, associate editor, was always available to answer questions and birddog some requirement. The editorial staff at St. Martin's Press turned to with a determination that matched my own, certainly, and without their thorough editing and leg-work in procuring photos, drafting maps, and providing me with additional reference material, this would have been a much lesser work and not nearly as timely.

Finally, I acknowledge the assistance of my wife, who encouraged me to take on this project, supported me while it was in progress, and acted as my initial editor, proofreader, and critic.

To all of the above, I am in your debt.

AUTHOR'S NOTE

This account of General H. Norman Schwarzkopf and his brilliant role during Operation DESERT STORM is an attempt to introduce the public to a general who has been inspirational in his performance.

It has been written not for historians or military analysts, but for the international community of mothers and fathers, spouses and children, siblings and sweethearts of the men and women of DESERT STORM.

The emphasis, of course, is on General Schwarzkopf's performance as the commander of the multinational forces that executed one of the most effective and cost-efficient battles in history. At the same time, the author has attempted to put the events of DESERT SHIELD/STORM in an orderly and easily readable sequence. I have deliberately made a conscientious effort to insure that the reader is not swamped with technical detail, charts and figures, lists of numbers and so on, although some such material must be included for a better understanding of the conflict.

There are times when the content of this book reflects the author's thirty years of professional military experience, and such content does not necessarily reflect the views of either the publisher or any of those involved in producing this account.

It was necessary to make certain minor assumptions based on my own experience; if there are any errors, they are mine exclusively and have been made in good faith. All quotes have been documented from other sources, and all events have been researched from numerous accounts. Where I encountered conflicting research items, I picked the one most compatible with my own military experience.

Certain passages reflect the extensive television coverage that served the public for the seven months of the crisis. And although there is some critical mention of the media within the text, the author wishes to acknowledge and thank those responsible for the overall extensive coverage of every minute detail, and the exhaustive reporting of the events of the war that provided to the public a running account of DESERT STORM as it happened.

While the hard-charging but lovable Bear is featured throughout this account, the magnificent military men and women of the coalition are also paid tribute in these pages.

With respect to U.S. forces, not only are our all-volunteer military services the most educated and highly trained in American history, they have been provided with advanced-technology weapons that are second to none. The troops went into combat under the leadership of an outstanding general and did so with the full support of their fellow countrymen. The allies fought beside them with equal enthusiasm and capacity.

Nevertheless, it was the ability of General H. Norman Schwarzkopf to meld all of his forces into a championship team, and his foresight in planning that met the exact threat Saddam Hussein presented. Allied forces followed with flawless execution the battle plan that many military analysts have declared to be singular in its completeness.

General Schwarzkopf enjoyed the trust of his coalition partners, his commander in chief, his civilian superiors, and the American people. He basked in the loyalty of his troops, and in return he gave them the greatest concern for their welfare. He gave new meaning to the word *blitzkrieg*—maybe we should even change it to *Schwarzkrieg*.

What else can you say of a man like General Schwarzkopf?

"Thank you," comes immediately to mind.

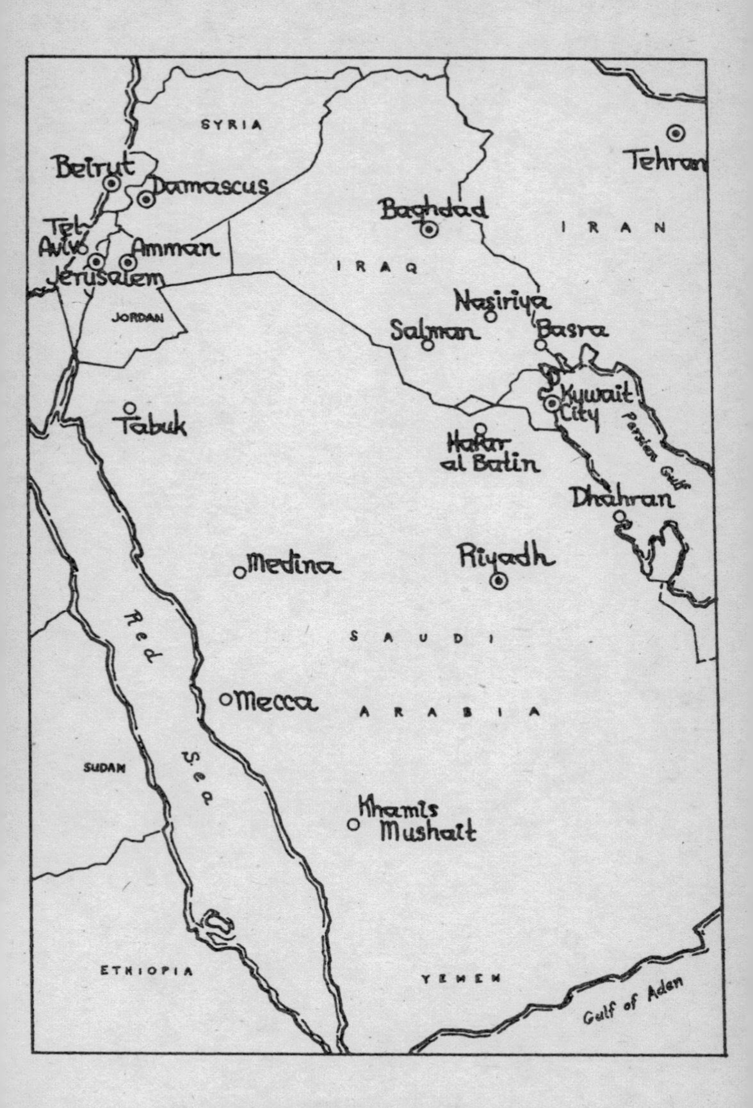

THE PERSIAN GULF REGION

FOREWORD

THE BIOGRAPHICAL ASPECTS of the story that you are about to read pertain to a remarkable soldier, a man cast in the mold of Grant and Sherman, Bradley and Patton and Eisenhower. He is both a rough-and-ready, self-assertive, infantry fighting man and a gentle, caring, loving human who in times past has played Santa Claus to his children and speaks of the horror and hardships of war and the separation from his family with the same deep emotion felt by the men and women placed under his care.

He is a soldier's soldier and a born leader who can not only exercise brilliant military leadership in the field but also handle the diplomatic and political aspects of top level command with sophistication and a constant concern for the interests of his nation.

He hates war and yet has had a lifelong ambition to lead his country's forces in a great decisive battle. As a reluctant warrior he is neither dove nor hawk, preferring to describe himself as an owl who hopes "to watch, to learn, and be wise."

To better understand the circumstances that have thrust him to the fore and provided him with his opportunity to serve his country in the most exemplary manner, a bit of background is in order.

Saddam Hussein set off the events that culminated in DESERT STORM with his invasion and annexation of Kuwait. His decision reflected a turbulent history of the region, including a previous threat by Iraq to invade the small sheikdom as far back as 1961.

Kuwait and Iraq were both originally part of the Ottoman Empire and the modern-day Kuwaiti royal family dates its origin back to 1756, when the first member of the Sabah family was selected to be the ruling sheik of the 'Anizah tribe who were, in a sense, the first Kuwaitis. Iraq at that time was already under the influence of Persia and the modern day Iraq-Kuwait border reflects the applicable portion of the old Ottoman-Persian border.

In 1898, the Turks had designs on Kuwait but were dissuaded by the British, the power in the area at that time. In fact, Britain reached out its long imperial arm and took Kuwait under its influence as a protectorate. Turkey recognized the autonomy of Kuwait just prior to World War I.

As for Iraq, during the "war to end all wars" Britain consolidated her hold on the gulf area by capturing Baghdad, and after the armistice, the League of Nations made Iraq a British mandated territory. Iraq became independent in 1932.

At the outbreak of World War II, urged on by the Axis powers, the Iraqis attacked British forces stationed in the area but were soundly defeated in one of the war's early allied victories. After the war, the British established Iraq as a monarchy and it stayed thus until the 1958 revolution. A number of coups occurred over the next thirteen years and the 1958 coup established the Ba'ath Party as the governing power, a position it occupied at the outbreak of the Allied-Iraq conflict of 1991. Saddam Hussein became the head of state in July, 1976, and shortly thereafter led Iraq into its eight-year war with Iran.

That long, exhausting conflict sapped Iraq of its wealth and Saddam Hussein found himself in a precariously fragile economic situation. Reaching back to the historic claim of Iraq upon Kuwait, he added the new dimension of OPEC's oil pricing policies over the past few years to come up with an arguable motive for invading Kuwait: the low price of oil had robbed him (i.e. Iraq) of billions of dollars of much-needed revenue, and Kuwait, as one of the principal oil exporters, owed Iraq its "lost" oil revenues. Saddam also knew that the acquisition of Kuwait would give him better maritime access to the Persian Gulf. These economic and strategic arguments could be used to hide a more insidious motive: the acquisition of a bigger chunk of the oil-producing deserts of the Middle East. With control of the oil fields of Kuwait, Saddam would have control of over 40 percent of the accessible world oil reserves. In a moment of vainglory, he must also have considered that if he could threaten and attack Saudi Arabia he might actually expand his control over as much as 70 percent of the planet's oil reserves! Iraq would be a world power with real influence over the world economy and Saddam Hussein would be the most powerful leader in the Arab world.

Accordingly, Saddam Hussein, the commander of the world's most formidable Arab military force, ordered his troops into the tiny and practically undefended sheikdom of Kuwait on August 2, 1990. Subsequently, he declared that territory formally annexed by Iraq and redesignated as the nineteenth province of that country. And he began to position his forces along the Kuwait-Saudi Arabia border for a possible further incursion into the world's richest oil fields. This was too great a threat for the world community to ignore.

And a very real threat it was! The average world citizen, accustomed to thinking of Saudi Arabia as a larger and more technologically advanced country than Iraq, was surprised to learn that in most military areas, Iraq—even after its manpower- and material-consuming war with Iran—was much more powerful than Saudi Arabia. Against almost a million troops, the Saudis had under arms only 65,700. There were only 550 Saudi tanks to counter an Iraqi force ten times that size. And the 189 combat aircraft of the Saudis would be defending against an attacking air force of almost 700.

In short order, the United Nations Security Council unanimously condemned the invasion, called upon Iraq to withdraw from Kuwait and voted to impose an embargo on all Iraqi trade outside of humane

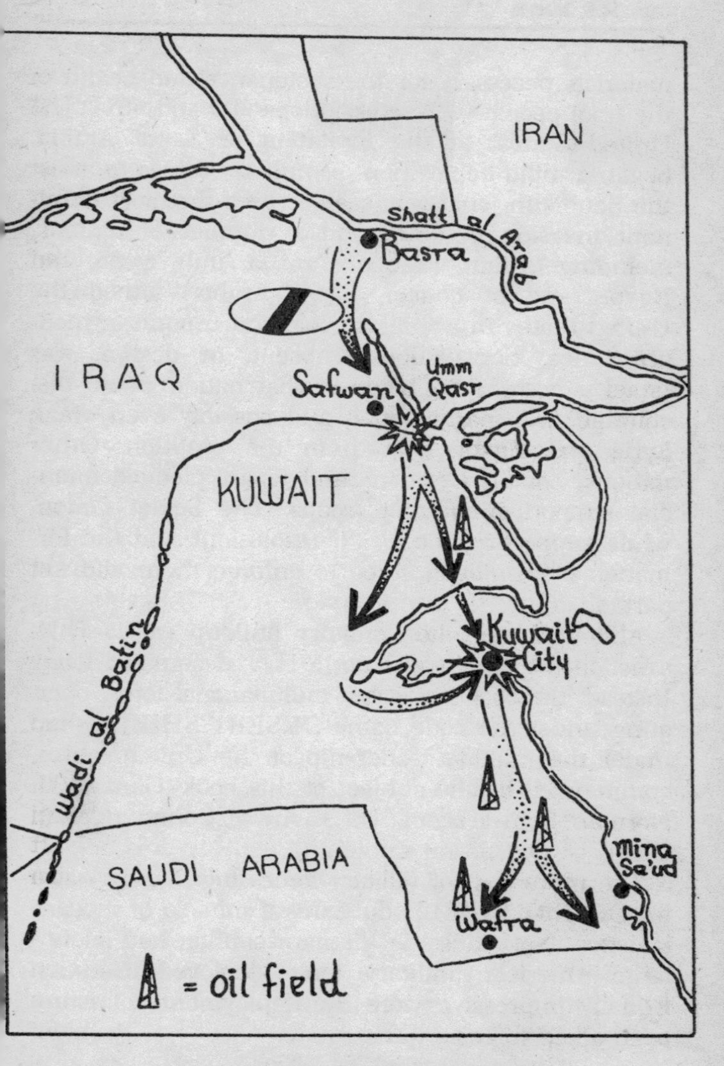

IRAQI INVASION OF KUWAIT—
August 2nd and 3rd, 1990

materials necessary for the sustenance and health of the Iraqi people (U.N. Resolutions 600 and 661). The United States, at the invitation of Saudi Arabia, began a rapid deployment of military forces to assist the Saudis in what was seen to be a threat of imminent invasion by Iraq; and a number of nations, including Britain, France, Canada, Italy, Syria, and Egypt (and of course, Saudi Arabia) joined the United States in a military coalition unique in modern times. Conspicuously absent, by design, was Israel, since the inclusion of that nation would discourage Arab participation and possibly even break Syria and Egypt away from the coalition. Other nations, notably Germany and Japan, pledged financial support in lieu of troops. The Soviet Union, while supporting the U. N. resolutions and the formation of a military force to enforce them, did not participate.

Although the allied military buildup was a U.N. sanctioned event, it was not a U.N. designated force. Instead, the coalition was a multinational force, operating under the code name DESERT SHIELD, and under the military leadership of the United States, commanded by the subject of this book, General H. Norman Schwarzkopf, U. S. Army, Commander in Chief, U. S. Central Command.

The movement of military forces into Saudi Arabia for DESERT SHIELD duty was a miracle of modern logistics. Not since the Vietnam conflict had such a large American military force deployed overseas. Equally impressive were the deployments of European allied forces.

With sea- and airlift resources that employed deployment routes equal to half the distance around the world, the United States led off with rapid-deployment airborne and marine troops. Air forces made the long journey with a minimum of stops, preferring to use in-flight tactical refueling capabilities to shorten the time necessary to get on station. Naval forces, some already on station in the Mediterranean Sea and Gulf waters, matched the deployment pace of their sister services and ever ready Coast Guard units responded with equal urgency. National Guard and Reserve units were placed on active duty and a number of these deployed. Others entered stateside training for desert warfare. It is a tribute to the planning and farsightedness of the past decade that such an extensive operation could be mounted with a minimal disruption of the American way of life. Forces were on hand. There was little talk of a draft and there was obviously no need for one.

There were some disturbing side effects. DESERT SHIELD/STORM would be expensive. The American economy, already in a slowdown, would have to shoulder a very untimely additional budget burden. Wartime demands on air and sea carriers would affect civilian travel (as would the foreseen terrorist threats). The uncertainties of wartime would cause the American public to rethink its priorities, and spending patterns would be affected; businesses would feel the effects of consumer caution and inevitably some would have to release employees.

The paper "peace dividend," created by the collapse

of Cold War financial requirements and anticipated as a shot in the arm for the domestic programs of the nation, simply disappeared.

There was the inevitable comparison with the Vietnam years and a rebirth of so-called peace movements. An extremely small number of military personnel decided that in all good conscience they could no longer serve in an organization where the killing of their fellow man would be required.

But the impending Persian Gulf conflict was in no way a repetition of this country's most unpopular war. There simply were no major similarities. The political and military aspects of the Persian Gulf situation were much more reminiscent of those that preceded World War II. The ideological struggle present between the two sides of the Vietnam conflict was not a factor. Instead, without warning or justification, an independent nation had been brutally and illegally attacked, conquered, and annexed by its aggressive neighbor in clear violation of the principles and charter of the United Nations. And the attacker was clearly poised to expand his conquest of the crucial oil-rich areas of the Middle East. The national interests of every nation that had to depend upon the import of oil for its vitality and progress were clearly at risk.

The world community was outraged as evidenced by the universal condemnation of Iraq and its leader, Saddam Hussein. *Never before* in modern times had there been such agreement between democracies, monarchies, republics, and even totalitarian governments. Saddam Hussein had overstepped the boundaries of civilized behavior.

And the people of the United States, as represented by their Congress and after a healthy discussion of the pros and cons of our political and proposed military actions, overwhelmingly supported the decisions of President Bush and empowered him to conduct military operations as he saw fit. Their approval was seconded by the United Nations when that body formally approved the use of military force any time after January 15, 1991, should political and economic efforts fail (U. N. Resolution 666). The nation and the world were united in their goal to force Iraq to withdraw from Kuwait. And the military forces that would be responsible for armed enforcement of that decision would not be restricted in any way. The military commanders in the field would fight the war. The lessons of Vietnam, as hard as they were, had been learned.

Still, in a final effort to avoid military conflict, President Bush proposed direct talks between himself and the Iraqi foreign minister and between Secretary of State Baker and Saddam Hussein. To no avail. Hussein, determined to exercise control over events, refused every proposed date for the talks, insisting that he would set the date and deliberately suggested ones so close to the January 15 deadline that such a meeting, even if it produced a past-the-deadline withdrawal, would constitute a face-saving victory.

Saddam's stubborn refusal to enter into any final negotiation and his insistence that the state of Kuwait was no longer in existence and instead was now permanently a province of Iraq was a most serious mistake.

By January 15, 1991, the forces assigned to DESERT SHIELD were essentially on station. The buildup was complete and Saddam Hussein was facing a military enemy which might be just a bit short of his forces in number but was decidedly superior in technical ability. In all probability, he had no real feel for the capability of the forces aligned against him. But that lack of understanding quickly disappeared on the morning of January 17 (Iraqi date).

DESERT SHIELD became DESERT STORM and from the skies over Baghdad came the first explosive signs of the beginning of the end for Saddam Hussein.

The architect of the brilliant campaign that enabled coalition forces to execute an epic air and sea assault, and to launch an historic ground battle that completely annihilated Iraqi military might in and around Kuwait in one hundred hours is the subject of this book.

CHAPTER 1

GENERAL H. NORMAN SCHWARZKOPF, United States Army, Commander in Chief, United States Central Command.

Commander, Operation DESERT SHIELD/STORM.

Stormin' Norman. The Bear.

America has a new hero.

Almost overnight, in terms of all-out warfare, General Schwarzkopf has led his superb forces in a desert battle that is destined to be taught and studied as one of the classic campaigns in military history.

Not only did his multinational forces defeat the fourth largest military organization in the world, they demonstrated a new solidarity of international cooperation and singleness of purpose that could well carry us into the next century with renewed confidence in a search for world peace. It may be that General Schwarzkopf should add the term Peacemaker to his many distinguished titles, for the next Saddam Hussein to come to the fore will think long and hard before embarking on a course of ruthless aggression. The thousands of shattered vehicles strewn across the Arabian desert are testimony to the fury of righteous men when they gather together to strike back at the forces of evil. War is a terrible thing and no one knows that more than the warrior. Yet, there are those among us who dedicate their lives to the science and art of war, all the while praying most fervently that they will never have to exercise their skill. It is one of those paradoxical realities that men have lived with since one of their kind threw the first stone at a fellow man.

H. Norman Schwarzkopf is just such a man of dedication. He is one of those rare individuals who has known his destiny since his earliest years and has guided his efforts toward that destiny with the passion and fears of the combat-hardened professional soldier. When he relayed the order "Execute DESERT STORM" from his commander in chief to the men and women under his care, he did so with the fullest personal knowledge of the gravity of that decision. To the complex conduct of military operations was added the final dimension: casualties. Some of those placed under his care and guidance would die. Others would carry the physical and mental scars of that order for the remainder of their lives. But all would know that they had acted with honor and with the most noble of purposes. They had offered themselves up for their fellow man. There is no greater act of love. Another paradox.

But General Schwarzkopf had a measure of comfort. He knew that the operation had been planned with careful consideration of all of the factors involved. He knew he had good deputy commanders and highly motivated forces. He knew that his people had the most sophisticated weapons in the world and the most intense training his five months of preparation could have given them. He knew that among his forces were those of other nations, allies who joined the United States in the conviction that the battle upon which they were embarking was just and necessary. He knew that he had done his very best in molding all of these forces into one united fighting machine. And when he gave his order, he was as confident as any commander can be that the thunder and lightning of DESERT STORM would achieve the goals with which they were charged.

A commander in the field must be many things but most of all he has to be an experienced leader of men. In that role he must possess qualities which will convince his troops that, while victory in battle is the primary goal, their well-being must be his personal concern. The troops of DESERT STORM knew that their general had been hardened by combat in Vietnam and from his own experience was well aware of their needs and fears. Schwarzkopf wanted to emphasize that, and one of the most common scenes of DESERT SHIELD was his helicopter swooping down to deposit him in the midst of his troops, to visit with them and share their field rations and listen to their comments. Of all the published pictures taken of General Schwarzkopf, the ones of him surrounded by his troops show him at his best. He is in his element, big arms draped across a pair of shoulders for a picture the young soldiers can send home to their families. Listening intently to their news of a new baby or a letter from home. Laughing with them at a battlefield anecdote.

There is a lot of telltale evidence of Schwarzkopf's love of his people in those pictures. He is not a man who smiles often when he is engaged in the serious business of war. But the upward sweep of his mouth when he is among his fighting troops sends a clear message. Here he is at his happiest, encouraging his men by his presence and receiving their admiration because they know he truly cares. It is probably the supreme example of male bonding, that of men in combat together. The foot soldier and the Old Man jawing together, for a brief time sharing the joy of soldiering. It is a rare moment and one that not all get to enjoy, but for those who do, it is a moment to remember and relate over and over in the times to come. It is one of those things that make up the soldier's greatest asset: morale.

Finally, while the general presents a macho front that would make John Wayne quake in his boots, he is a gentle, sentimental man whose eyes well up with tears when he speaks of the sacrifices of his men, his love of country, the separation from his family, and the fond memories of his father.

The Persian Gulf War has given new life to an old and somewhat bizarre theory. There is one obscure school of thought that theorizes the existence of a duplicate universe in another dimension and in that universe we find the exact opposite of our own reality. Where there is light in our universe, there is darkness in the other. Where there is black and white, we have white and black. Where there is a saint, we have a sinner, and where there is a sinner we have a saint.

Most of us have little interest in such a far-out hypothesis but suddenly we may have dynamic evidence that supports the "theory of opposites." The events that occurred in the Persian Gulf area between August 2, 1990, and February 27, 1991, have given a mighty credence to the idea that for every evil there must be a good.

Because while the dark universe has Saddam Hussein, we have General H. Norman Schwarzkopf.

Never are our senses of freedom and justice and world responsibilities so aroused as when our beliefs or national interests are threatened. Invariably from within our ranks there rises one or more of us who is equal to the times and sets the example for all who would defend our liberties and way of life.

From World War I with John H. Pershing and a young Douglas MacArthur, to World War II with George Patton, to the Korean War with Dwight Eisenhower and an older, determined MacArthur—great men have risen out of devastating times.

Then came the horror of Vietnam. For the first time in our history, the civilian authorities pre-empted military strategic and tactical goals down to platoon and individual aircraft and ship level, and the military found itself in a position of such apparent ineptness that it bordered on the dishonorable. It wasn't a dishonor, of course, for the U.S. military fought as bravely as they had in any war. They just could not achieve the preferred military objectives. As one soldier in the field was reported to have said, "Johnson may be a great president but he's a piss-poor squad leader." This is no place to debate that war—that's for many others to argue—but the conditions during the Vietnam conflict did not favor the emergence of a great warrior. America needed a great military leader but had to be satisfied with civilian "management." You "manage" a 7-Eleven convenience store; you *lead* men into battle. But at least, a great warrior for the future was being shaped by the horror and frustration and tragedy and disappointments of the Vietnam battlefield: H. Norman Schwarzkopf.

Schwarzkopf had that precious quality that is one of the hallmarks of a great warrior and military leader. He loved his men. After an initial stint as adviser to South Vietnamese paratroopers, he commanded an infantry battalion, finally returning home in 1970 as a lieutenant colonel. To see his troops bleed and die under the conditions they were thrust into must have angered him to the point of rebellion.

Like so many other soldiers, Schwarzkopf has very unpleasant memories of 'Nam, referring to them as scars that will never go completely away. He knew the vast majority of his comrades had fought well there without the support of much of the news media or a large segment of the U. S. population. The memory that fighting men were called baby-killers and spat upon still causes his eyes to flash and the hair on the back of his neck to bristle. He will never forget the tragic time when he was a battalion commander and one of his companies had lost its commander and its lieutenant. The company was stranded in the middle of a minefield. Schwarzkopf, along with his artillery liaison officer, Captain Bob Trappert, who had been overhead with him in the command helicopter, had the Huey land, and they took charge. He and Trappert started to lead the men out of the dangerous area, back the way they'd come. Suddenly, a trooper stepped on a mine. Horrified, but with the presence of mind that made him a true battlefield commander, Schwarzkopf threaded his way through the minefield to the wounded youngster, calmed him, and tended to his wounds. Then, tragically, Trappert stepped on a mine. He was gravely injured (but subsequently survived) and three others were killed. Schwarzkopf finally led the survivors to safety. One of those Silver Stars on his chest is a constant reminder of that nightmare.

And it was in the agony of battle that the temper of H. Norman Schwarzkopf made its first appearance. In an interview with reporter Joseph L. Galloway, Schwarzkopf stated, "The first time I ever really lost my temper was in a situation at Duc Co. I had wounded Vietnamese lying on the ground who desperately needed air evacuation, and I had helicopters flying all over the place with VIPs in them and I couldn't get the helicopters on the ground to get the wounded out of there. I can still remember it to this day because I was yelling profanities over the airwaves. I can remember subsequently getting back to the States and suddenly I would lose my temper. How shocked I was; it had never happened before. I want to say that I do not get mad at people; I get mad at things that happen. I get mad at betrayal of trust. I get angry at lack of consideration for the soldiers. And contrary to what's been said, I do not throw things. If somebody happens to be in my burst radius when I go off, I make very sure that they understand it's not them I am angry at. Having said that, anytime a guy who's six-foot-three and weighs 240 pounds and wears four stars loses his temper, everybody runs for cover. I recognize that, but I don't think I'm abusive. There's a difference."

As disillusioned as he was, Schwarzkopf fought in Vietnam with a warrior's code that had been drilled into him at one of the schools of great warriors: West Point, Class of '56. And that code was Duty, Honor, Country. Consequently, when he left Vietnam, he did so with three Silver Stars and two Purple Hearts in places of honor among the multicolored ribbons that were arrayed just below his Combat Infantryman's Badge and Jump Wings. Also, as a constant reminder of those days, he carries around a recurring back pain, a souvenir of the physical demands of earning and wearing paratrooper insignia.

Duty, Honor, Country. It's a demanding code and all of the military services have identical codes although they are stated perhaps differently. It's a code that says one hell of a lot in just three words. And it is an uncompromising code. You don't get to pick two out of the three.

Whenever one goes back and studies the great men of military history, one almost invariably finds that they seem to be predestined. General Schwarzkopf has Duty, Honor, and Country imprinted in his genes. His father, Herbert Norman Schwarzkopf, was also a member of the long gray line, a West Point graduate, and a very successful career soldier who, despite a break in his service just prior to WW II when he pursued a civilian job in law enforcement, earned his stars prior to retirement.

As if his son's destiny were guiding the father's career, the senior Schwarzkopf spent the war years and a time afterward (1942–1948) as the head of a military mission to Iran. There he learned the ways and cultures of not only the Persians but of the Arabs who were Iran's neighbors. Young Norman had accompanied his father to Teheran as a twelve-year-old and experienced life in that then remote part of the world.

All of us are products of our environment. But we are also more than that, or less, in too many tragic cases. We are as much the product of our own motivation and initiative. Everyone has goals, although these goals may change as we go through life. Not so with H. Norman Schwarzkopf (Just a plain H. His father was not very enthusiastic about his own name, Herbert, but the H would carry on the lineage). H. Norman seemed to know from his earliest days that his goal was the military and in his young naive mind (or maybe it wasn't so naive), he believed he would be a general like his father. As a pre-teenager, he entered the Bordertown Military Institute near Trenton, New Jersey, the place of his birth. He spent time in Iran with his family. He received further education in Europe and early on was exposed to two languages that have considerable military relevance, German and French. His family background, his diverse schooling and his excellent performance in school, as well as his demeanor and mindset, all led to the winning of the coveted appointment to West Point. There, he enthusiastically entered into all of the activities (formal education, military training, and body-conditioning sports) with a singleness of purpose: to be the best. He even found time in his senior year to join and direct the choir and is quoted as having later said that the difference between directing music and troops is that in the latter case "the orchestra starts playing, and some son of a bitch climbs out of the orchestra pit with a bayonet and starts chasing you around the stage."

It was only natural that his dedication, observed by his classmates, earned him the nickname of Stormin' Norman, a moniker he wears to this day and one that has its own ring of prophecy. He is not crazy about that name, believing it to be descriptive of someone who charges ahead willy-nilly without regard to the consequences. To most of us it means just the opposite and we speak it with affection. It reflects his dynamic leadership, his nonwavering intention to do his very best; and his determination to serve his country.

With that definition in mind: Who better to lead a DESERT STORM than Stormin' Norman!

General Schwarzkopf is 56 years old, six-foot-three, has an IQ of 170, and reportedly weighs 240 pounds as of this writing. When you see him face to face, he is an impressive and intimidating gentleman. He doesn't strike you as particularly overweight, despite ample jowls. Rather, he strikes you as the type of man that you would dearly love to have on your side in a dark alley or a barroom brawl. His alternative nickname, the one he prefers, The Bear, is equally descriptive, both from the standpoint of his menacing physical bulk and his reputation for having something of a temper. This latter trait is possibly misunderstood by some of those who have incurred his wrath or written about it with some disdain, for a military leader must have a temper. How he controls it is another matter, and one perhaps subject to interpretation, but he must be willing to use it any time the less-than-best performance of a junior or the inadequacy of a plan threatens the lives of his men or the success of an operation. When it comes to the awesome responsibilities of a general in the field, there are times when a little anger can be a good thing. And with Schwarzkopf, his displays of temper seem always to be followed by an encouraging word or a conciliatory comment.

Women very likely see him as gentle, certainly handsome enough in a mature way, and intriguing. Those dark eyes—that to a fellow soldier project laser beams demanding performance—probably present a challenge to the fairer sex. Here is a man they would like to tame, a big cuddly teddy bear who can be a gracious escort, enjoys the sophisticated strains of operatic arias, and carries the hint of a smile most of the time on a face that must come in very handy during poker games.

Inevitably, we compare General Schwarzkopf to previous prominent military leaders. Patton immediately springs to mind, but on second thought there appear to be more differences between the two than similarities. For all of his military genius, Patton was flamboyant, a bit crude, sometimes impatient, and had an annoying habit of putting his foot in his mouth. Schwarzkopf appears more self-assured and quietly confident. He is cautious in his statements and saves his expletives for the rare times when their impact is needed. He is diplomatic in his public statements.

Patton loved the spit and polish of a general officer's military uniform and took great pains with his image which was often inspiring to his troops. Schwarzkopf is equally inspiring but prefers the foot soldier's desert battle dress. Patton wore twin ivory-handled revolvers; Schwarzkopf seems to prefer twin watches, one on each wrist, perhaps to keep track of Washington time as well as that of the theater of operations.

But the two *are* similar in the most important things. Both have displayed their skills as brilliant strategists. Both have practiced the art of military planning in a way that provides the maximum probability of achieving their objective with a minimum number of casualties. And both have displayed a great love for their men.

In the movie *Patton*, there is a poignant scene where Patton comes upon the devastation and debris of a recently fought tank battle and, amid the torn and bloody corpses of both sides, there is one survivor, a severely wounded American sergeant sitting in the sand with his back resting against the remains of an armored vehicle. Patton (played with superb conviction by actor George C. Scott) drops to one knee and speaks softly to the sergeant, then removes his own helmet and leans over to gently kiss the fallen soldier's forehead. The scene is most probably fictitious. Even if it is, it makes its point well. Patton loved his men with a passion. Each one was special to him and he grieved with each loss. Why? Because he, too, was a warrior and he knew from his own experience the fear and bravery of men in combat.

Such a man is General Schwarzkopf and as we begin this profile with emphasis on his obligations and performance during DESERT SHIELD/STORM it would be wise to take a few minutes and look at the significance of the four stars that mark his attainment of the highest rank in the U. S. military.

Those stars are not awards. Nor are the Combat Infantryman's Badge and Master Parachutist Badge above his left pocket. They are indicators of exceptional military achievement. Take a look at any general or flag officer, regardless of the branch of service. Almost universally they wear either those badges or a set of wings or submariner's dolphins or surface warfare insignia. Each emblem testifies to that officer's ability to complete the most rigorous training or combat requirements and each of those badges has been earned in an environment that always resulted in one or more of that officer's companions being killed during the actions required to earn that badge. And lest we forget, several of those badges can be won by enlisted men as well as officers and the requirements are identical. Yet while such qualifications are not necessary for every soldier, sailor, marine, or airman, they are crucial for those who at the highest level would lead men in battle.

Every general starts off as a second lieutenant (there have been exceptions but very rarely). Promotion to first lieutenant is almost routine (a 98–99 percentage being typical) but there are a few who are judged not qualified. Advancement to captain is well within the capabilities of most of the first looies but the percentage who fail rises just a bit, perhaps five percent. But now the numbers game begins to play a factor. There can be only so many majors; thus forced attrition may remove some very fine officers who would otherwise be capable of increased responsibilities. The jump to lieutenant colonel becomes a significant hurdle with only the very best (perhaps 50–70 percent of the majors eligible, depending upon force requirements) being promoted, and those who are selected represent the most professional of the lot. To colonel, the percentage drops to typically 40–60 percent of eligible lieutenant colonels, and those who make that cut are the best of the best in performance (they will normally have already displayed command-in-the-field ability) and career potential. Promotion to brigadier becomes not only a matter of performance but of career reputation and the difficulty of previous assignments (some assignments are more demanding than others). It also routinely represents the completion of postgraduate education and advanced formal military training (typically a service war college—Schwarzkopf attended both the Army Command and General Staff College and the Army War College). By the time a soldier pins on a star he has competed—over and over—against the best military personnel the United States can field. Two stars is a further competitive achievement but few two-star generals are selected for three, primarily because of congressional limits on the number of O-9 officers allowed (O-1 being the second lieutenant rank designator). Thus, a large percentage of very superior officers must of necessity be passed over. Selection for the rank of full general becomes a matter of individual scrutiny by the highest levels of the military and DOD organizations and normally involves input by the commander in chief, the president of the United States. Obviously, the exceptional officers who achieve four-star rank have all proven themselves against the very highest of standards. They represent much less than one half of one percent of those who started along the road, and they have typically given more than thirty years of their life to achieve that rank (author's note: the figures used are not necessarily exact due to changing requirements over a thirty-year career span, but they are sufficiently representative to make the point).

By the time an officer pins on a fourth star he has experienced every facet of his chosen career and has demonstrated command ability at the highest level of operations.

The factor, then, that sets one four-star general apart from the others is his personal modus operandi and some special quality that has been displayed during periods of crisis. And it takes nothing away from General Schwarzkopf to point out that the military of the Western powers are blessed with a number of general and flag officers who can rise to similar occasions. American officers are products of a rigorous system and one can have every confidence that it is the best military system for recognizing ability and achievement among the many nations on this earth. It was *expected* that General Schwarzkopf would perform according to the highest tradition of the military service. It is to his everlasting personal credit that he exceeded those expectations! His personal operating philosophy and ability to meet every command problem with a solution have raised him just a bit above his four-star rank. He would be the first to give credit for his achievements in the Persian Gulf to the staffs and operating forces of all the international services under his command but we cannot escape the fact that his ability to meld those forces into the most efficient fighting machine in modern times has made him an authentic hero.

But even a military hero has to be in the right place at the right time. So, how did General Schwarzkopf find himself in a position to fulfill his lifelong ambition—it has been reported that as far back as his days at West Point, he had stated that not only was he going to be a general, he was going to lead United States forces in a very significant battle. To begin with, it started the first day he pinned on his gold bars as a second lieutenant. The most junior of officers have little control over their destinies at that point in their careers, but there was one factor over which Schwarzkopf had complete control: his performance. From the earliest days, Schwarzkopf has been the consummate soldier. Though his responsibilities were limited then, he had one overriding standard for meeting those responsibilities: be the best. It is interesting to point out that the U. S. Army's current slogan, "Be all you can be!" is a lot more than recruiting propaganda. It is a very real reflection of the attitude that makes these professional soldiers among the best in the world. From dogface to general, being the best you can be is a sure path to achievement and satisfaction. Schwarzkopf may not have originated the thought but it certainly represents his frame of mind. And he is rightfully proud of the fact that as a mud-foot infantry officer he has gone all the way from second lieutenant to four-star general. It would have pleased him greatly if his father could have lived to witness his career.

The military path to the level of general officer is fairly standard from the standpoint of assignments. Platoon leader, company or battery commander, battalion commander, regimental commander, brigade commander, and division commander are all rungs in the steep ladder of an army officer's career. Not all make every rung; most stop short of the top and are shunted off to other duties. A few actually skip a rung or two but continue on upward and even command a corps. By then they have earned three stars. The rare ones that don the fourth star may command an army, the largest military field unit, which normally consists of several corps and a number of divisions.

In between field assignments, an officer must fill a number of staff positions and further his formal education and military training by earning advanced degrees (often in his "spare" time) and successfully completing courses at one or more of the service war colleges.

And finally, when an officer finds himself at the top of the ladder, he may be assigned as a commander in chief of one of the largest and most complex military U. S. commands, the Joint Command, a unified force that contains assigned elements of all three services and is the most senior operational military command within the Department of Defense.

In his position as Commander in Chief of the United States Central Command, General Schwarzkopf is directly responsible to only one man, the Chairman of the Joint Chiefs of Staff (JCS), and as fate would have it, General Schwarzkopf's operational commander during DESERT SHIELD/STORM has been one of the finest and most capable military leaders the U. S. military system has ever produced: Colin Powell, U.S. Army.

General Powell, at age 53 the youngest officer ever

appointed to the position of Chairman of the Joint Chiefs, is also one of the heroes of our day. A two-tour veteran of Vietnam with a most distinguished career that has included significant combat and command assignments, all performed with distinction, General Powell brings to the highest military office both military and political expertise—a rare combination, but exactly the right mix to direct the duties of General Schwarzkopf and advise his own senior, the Secretary of Defense, on military matters. Amazingly enough, General Powell is not a graduate of West Point. He is a product of the Army's ROTC (Reserve Officers Training Corps) with undergraduate work at City College of New York. Indicative of his exemplary career achievements is the fact that he was picked over fifteen more senior generals to ascend to the position of Chairman of the Joint Chiefs.

The son of immigrant Jamaican parents who spent his youth in Harlem and the South Bronx, he is a notable witness to the opportunity that is available within the military to Americans from all of the various social strata in this country. In addition to his combat and operational assignments, he has served at several posts within the White House, having been recognized as a "comer" by Reagan's Secretary of Defense, Caspar Weinberger. He served as Deputy National Security Adviser to Frank Carlucci during the Reagan administration. Consequently, he is acutely aware of both the obvious and the subtle interplays between the highest civilian level of the government and the political as well as the military nuances that are always a factor in United States military decisions and operations. He is as well regarded for his political astuteness as for his military achievements although he does not enjoy being labeled a political general. That is understandable, for he is demonstrably a military leader in a position that has critical political aspects.

The position of the Chairman of the JCS has changed since the early days when the role was more of a spokesman for the senior service chiefs, rather than the top military leader. In 1986, the office was given more power over the Joint Chiefs and his own staff increased authority. He has the reputation of being a good listener and a good decision-maker.

General Powell knew that the commander must have the flexibility to plan and conduct operations according to the conditions that exist in the field. He had experienced the anguish and futility of being micromanaged by upper echelons during the war in Vietnam and had every confidence in Schwarzkopf's ability to conduct the battle. It would be Powell's job to approve the plan of operations and support Schwarzkopf by making certain that all of the required assets were available to carry out the operation.

He had another equally important responsibility—to provide the Secretary of Defense (SecDef) with military advice and counsel and provide the vital link in the command chain between the civilian authority and the military commander. On Powell's shoulders would rest the dual responsibilities of providing SecDef with military options to assist the President in his decision-making process, and insuring that the military was ready to execute the orders of the President.

The third member of the trinity (Commander Central

Command, Chairman of the JCS, and SecDef) is the civilian director of the Department of Defense, Dick Cheney. Once again, the United States had in its service a man of superb ability who understood fully his role in the management of the armed forces and his responsibilities to the president. He runs the department with a firm hand and is not at all bashful about responding to conduct that he feels is detrimental to his authority. When a high ranking Air Force officer took a matter to members of the Congress without informing Cheney, he received a severe reprimand—and in public. And when Air Force Chief of Staff General Michael J. Dugan publicly indicated that the Air Force would target Saddam's wife, children and mistress, Cheney relieved him immediately and the general retired.

Cheney has every confidence in Colin Powell and gave the general his full support. Tough in his intent to see that the military operates within the fiscal and manpower limitations imposed by the president and the congress, Cheney nevertheless supported every request by Powell for forces required to carry out Schwarzkopf's plan. In addition, Cheney undertook the difficult diplomatic task of assisting the President in convincing the nations of the world to join the United States in its efforts to reverse the invasion and occupation of Kuwait by the Iraqis. And the response to the crisis was, for all practical purposes, unanimous, and for the first time a military coalition was formed that found the peoples of the West and the Middle East united in their intent to right a terrible wrong. Those who did not send troops pledged to share the costs (Germany's and Japan's constitutions forbade them sending forces, for example).

As war clouds began to creep across the desert horizon, Schwarzkopf's joint command became multinational and the world watched as his genius became evident.

CHAPTER 2

ON AUGUST 2, 1990, General Schwarzkopf was at his Central Command Headquarters (USCENTCOM) at MacDill Air Force Base, Tampa, Florida. He was well suited for his duties as Commander in Chief (CINC) of one of the country's most critical commands. His career had included command at every level from platoon through corps, including command of two infantry brigades, command of the 24th Infantry (Mechanized) Division, and Deputy Commander of the Joint Task Force that carried out the Grenada operation. Later he had commanded I Corps and Fort Lewis, Washington.

In addition, the general had served as the Deputy Director for Military Plans and Policy of the United States Pacific Command, and Director of Military Personnel Management and Assistant Deputy Chief of Staff for Personnel, as well as Assistant Deputy Chief of Staff for Operations, Headquarters, Department of the Army.

Most recently, he had served as Army Operations Deputy, Joint Chiefs of Staff, and Deputy Chief of Staff for Operations and Plans, Headquarters, Department of the Army.

On the same date, halfway around the world, Saddam Hussein was in Baghdad, receiving the first news about his troops crossing the border into Kuwait, secure in his thoughts that he was embarking on a road that would propel him to the position of leadership within the Arab world. At one time, before he made his move into Kuwait, he had probably thought that such an action would inevitably lead to war and he was prepared for that. Even if he lost militarily, many reasoned, he could still come out a political winner. After all, in 1967 Nasser of Eygpt had been militarily defeated but emerged as an Arab leader.

Possibly Saddam believed his own writing, for in 1977 he authored a book titled *Our Struggle* and it was published in Germany (*Unser Kampf*—sound familiar?). In it, he described his intent to unite the Arabs under his leadership, destroy Israel and give the land to the Palestinians, and use an innovative oil policy to break the United States away from Japan and Europe.

He expected a worldwide reaction to his incursion into Kuwait but he knew there was no Arab power or combination of Arab powers that had the military ability to challenge his actions. He would invade, occupy, and then annex Kuwait in the fulfillment of ancient claims.

With Kuwait and its great wealth and resources firmly in the service of Iraq, he could begin to climb out of the economic hole he had dug during his eight years of war with Iran. It was entirely possible that acting from a position of such strength he could also threaten Saudi Arabia and gain control of a large chunk of that country's oil resources. To that end, he would position his forces along the Kuwaiti-Saudi border once they were firmly in control of the tiny Arab emirate. The only unknown of any concern to him at that moment was the reaction of the United States, but his evaluation was that the U.S. public would not support any militant action by their country even though their national interests were involved.

In Saddam's mind, the Americans were still soul-searching and divided over their actions in Vietnam, and their later invasions of Grenada and Panama had been clumsy efforts against minimal enemy opposition. The American people would not sanction military action against the great forces of Iraq. The memory of Vietnam losses was still too vivid. It would take a great war to dislodge him, and the Americans would have no stomach for that.

The immediate response of the United States and the world community was economic sanctions and condemnation of his actions. However, on August 4, 1990, President Bush took the first steps to order the first American troops to Saudi Arabia and militarily assist the Saudis in their defense against any further incursions by the Iraqis.

America's traditional friends Britain, Canada, and France started initial deployments of air forces. West

Germany, Canada, Australia, the Netherlands, and Belgium all sent naval forces to the gulf to assist in the sea embargo. The Arab League sent troops. An incensed world was showing unusual unity in response to the recklessness of Saddam Hussein.

The Iraqi leader countered with bombastic and irresponsible outbursts. Iraqi forces were invincible.

He had no way of knowing that the Bear was way ahead of him.

The United States Central Command is the administrative headquarters for U.S. military affairs in eighteen countries of the Middle East, Southwest Asia, and Northeast Africa including the Persian Gulf Region. It was established by President Reagan in January, 1983, as the evolutionary successor to the Rapid Deployment Joint Task Force.

Its mission is to support free world interests in its Area of Responsibility (AOR) by assuring access to Mideast oil resources, helping friendly and regional states maintain their own security and collective defense, maintaining an effective and visible U. S. military presence in the region, and deterring threats by hostile regional states as well as protecting U. S. military forces in the AOR.

USCENTCOM's AOR is geographically larger than the United States itself, encompassing eighteen countries of diverse political, economic, cultural, and geographic makeups. Three of the world's major religions have their roots there: Christianity, Judaism, and Islam. The topography includes mountain ranges with elevations of over 24,000 feet, as well as desert areas below sea level. Temperatures within the AOR range from below freezing to over 130 degrees Fahrenheit during the summer desert days.

In the center of the AOR are located more than 70 percent of the world's oil reserves, upon which the economies of the United States and its allies heavily depend. The AOR also sits astride the major maritime trade routes that link the Middle East, Europe, Southeast Asia, and the Western Hemisphere.

True, at that particular moment, Schwarzkopf was commanding largely a paper force. Along with five other geographical U.S. commands, the Central Command had few permanently assigned combat forces (the others are U.S. Forces Command in Georgia, Combined Forces Command in South Korea, U. S. Army Europe in Germany, Pacific Command in Hawaii, and Southern Command in Panama). But in the two years since reporting on board, the barrel-chested general had toughened his 700-person tri-service staff with a stiff injection of The Bear's personality. Their tasks had been to prepare and keep updated various contingency plans for consideration by the Joint Chiefs should a crisis arise within USCENTCOM's Area of Responsibility. And a feature of USCENTCOM's contingency plans was always rapid deployment.

A military contingency plan is the commander's proposed reaction to a specific threat, and contains certain assumptions about what has generated the need for the plan. Often politically sensitive, it is highly classified, and dissemination is made on a strict need-to-know basis. Once completed, it is sent to the staff of the Joint Chiefs for approval. It is then maintained on file both at the JCS and originating command level (along with a number of other such plans) and updated periodically.

The staff that prepares the plan includes officers of the three U.S. services who are responsible for their own service's input to the plan including the determination of forces required. The forces are not assigned to the proposed operation until the plan is activated, and it is not unusual for a specific military unit to be listed in a number of contingency plans as a force to be assigned. Normally, this presents no problem as contingencies tend to occur one at a time.

The contingency plan also contains provisions for strategic and tactical operations, communications, command and control, logistics, and other requirements such as psychological warfare, search and rescue, and special operations.

Thus, when an anticipated problem materializes, the commander responsible for meeting the threat already has a plan of action and his response can be that much more prompt. And it would take only one word from the president, "Execute!" to create a force in being and start the deployment.

General Schwarzkopf, having studied USCENTCOM's area and considered the possible threats had some time earlier ordered a contingency plan prepared for a possible U. S. response to an invasion of Kuwait by Iraq and had just completed a computerized war game (CPX, Command Post Exercise) of the plan in July! It gave him some valuable insight but he knew that an actual operation could and most probably would differ considerably from any plan or war gaming scenario. But it was a start and it generated serious staff thinking.

Schwarzkopf had also studied Saddam. He knew of Saddam's compelling ambition to take the position of leadership of the Arab world. He knew Saddam to be a ruthless leader, a man to whom torture and murder meant little, a man to whom the use of any weapon was justified by his ambition. Saddam was a known assassin and had used poison gases against fellow Iraqi citizens, the Kurds. The man seemingly had no concern about the suffering of his people that resulted from his actions. Schwarzkopf also felt that Saddam considered himself a man with a destiny predetermined by Allah, and such a man was dangerous. As for Saddam holding the rank of Field Marshal, Schwarzkopf was not impressed and would have a few words to say about that at a later date. He did respect Saddam's military might, however, and he knew that in his planning he would have to be prepared to fight in a worst-case scenario.

At one point after assuming his duties at USCENTCOM (on November 23, 1988), Schwarzkopf had considered retiring. He had served his nation long and well and gone through the hell of Vietnam. But now, the great battle that he had dreamed of in his youth was a possibility. He did not relish the call, for he knew it meant death and destruction. He hated war with a passion and his dream of leading a great decisive battle was certainly not predicated upon glory. Far from it. His motivation was love of country and his desire to be there when he was needed. He knew he had experience and talent; he knew he had conviction; now he knew he had purpose.

There was another very private thought. Schwarzkopf was a reluctant desk soldier. He always had been. During his second tour in Vietnam, when he was assigned as Battalion Commander, 1st Battalion, 6th Infantry, 23rd Infantry Division, he had constantly sought opportunities to be away from headquarters and get with his men. He knew the importance and necessity for staff and paperwork, but now he might be going back into the field! If war came, he was once more going to lead troops in battle. Amid all of the apprehension and second thoughts about the awesome responsibilities he was undertaking, there was one tiny little bright spot that showed through his bear-like scowl as a twinkle in his eyes. He was back to soldiering!

Pre-CENTCOM retirement thoughts were put on the back burner. His wife (the former Brenda Holsinger) and children (Cynthia, 20; Jessica, 18; and Christian, 13) were uppermost in his thoughts as he made that decision, and it was painful to recall how he had missed family life while in Vietnam. He and Brenda had been married after just a year of courtship. She had been a TWA flight attendant and they had met at a football game in 1967. They had endured sixteen moves together and numerous separations. For a loving husband and father, another deployment would be a miserable time. But he would be sharing that separation with hundreds of thousands of his men and women and their hardships would be much more severe than his. That made it both required and acceptable.

Even as he started taking his first stateside steps to generate DESERT SHIELD, the plan to deploy to Saudi Arabia and set up an immediate defensive line to prevent Iraqi forces from carrying out any designs they might have on Saudi oil fields, he knew that his superiors were going about their responsibilities. You just couldn't up and deploy a large military force to a foreign country without an invitation. Schwarzkopf knew he was going halfway around the world, and the Saudis knew he was coming, but the formal invitation had to be extended. The diplomatic process had to be completed and it was with some urgency that Dick Cheney, acting upon the orders of the president, visited King Fahd in Saudi Arabia and the invitation was extended.

Schwarzkopf and Company were ready, and on August 7, 1990, the deployment started.

The U. S. Central Command, as all joint commands, consists of a headquarters and three service component commands. Basically, the block diagram looks like this:

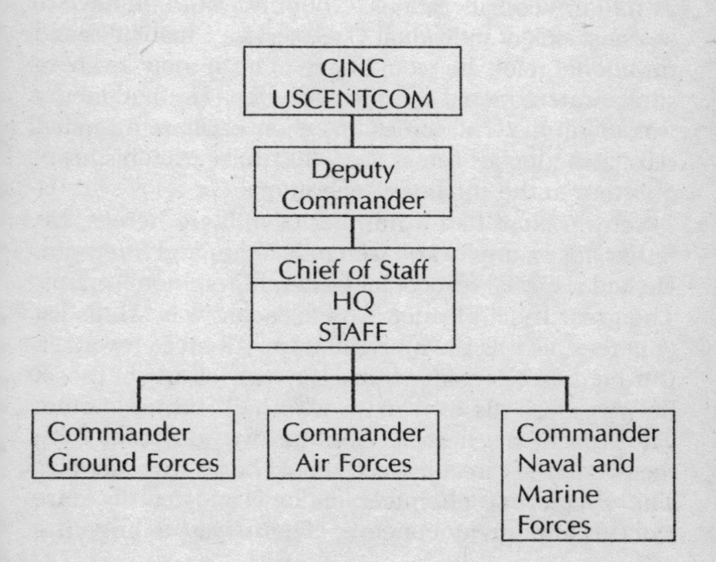

But Schwarzkopf realized that his stateside basic command structure was generic and only something that he would have to add to, as it would change with the addition of troops from other nations and his own concept of operational control. He anticipated the assignment of Arab troops, certainly from Saudi Arabia and Egypt but perhaps also from the smaller Arab nations. There were sensitive issues of national sovereignty involved. Traditional allies would also be sending forces, primarily the British and French and Canadians. Organizing coalition sea and air elements should present few serious problems, but when the ground phase began, the specific deployment of multinational troops and troop commanders would involve political as well as military considerations. Not only would he have to be conscious of individual U. S. service capabilities and traditional roles, he would have to be acutely aware of similar international considerations. But his first priority was to get his U.S. forces and their equipment, materials, and supplies across the 7,000 miles from his headquarters to the theater of operations.

Schwarzkopf had a number of military heroes, his father, for example, and Generals Grant and Sherman. He had a special respect for General Creighton Abrams. The great Indian leader, Chief Joseph, was on his list of heroes, as was the humanitarian, Albert Schweitzer. But he didn't consider himself a copy of any of those; he was more his own man, who had learned certain values from his heroes. His time was different from theirs and the threat he faced could be far more serious. The specter of chemical and/or biological warfare caused him great concern. There was a lingering thought about how the war had been micromanaged in Vietnam. He knew those days were over and that there was a completely different political and military atmosphere, but there would be pressures on him. But among all of his doubts there was one strong reserve. He was going to fight this war—if it came—based on his own evaluations of the threats and circumstances. He would not be pressured into taking any course that he did not feel was militarily sound. And one reassuring thought was the knowledge that his direct commander, General Powell, was a man of similar disposition.

There was another thought. Schwarzkopf was confident his forces would win any battle although he was unsure of the price to be paid. He prayed it would be minimal and it was his first intent to do all within his power to make it so. But he also wanted to conduct and win the impending war in a manner that would support peace and stability in the region. To all of those ends, he began to consolidate his headquarters in Riyadh, react to the buildup of coalition forces with precise command and control decisions, and train his forces in the sands of Saudi Arabia.

The elite 82nd and 101st Airborne Divisions had been the first to deploy in early August along with the 24th Mechanized Infantry Division and the 11th Air Defense Artillery Brigade and they were soon to be joined by U. S. Marines. One of Schwarzkopf's first concerns was their billeting and supply. It was the middle of summer in the desert and water alone was a significant requirement.

There was some prepositioned equipment and the first stages of a massive sea and airlift were under way.

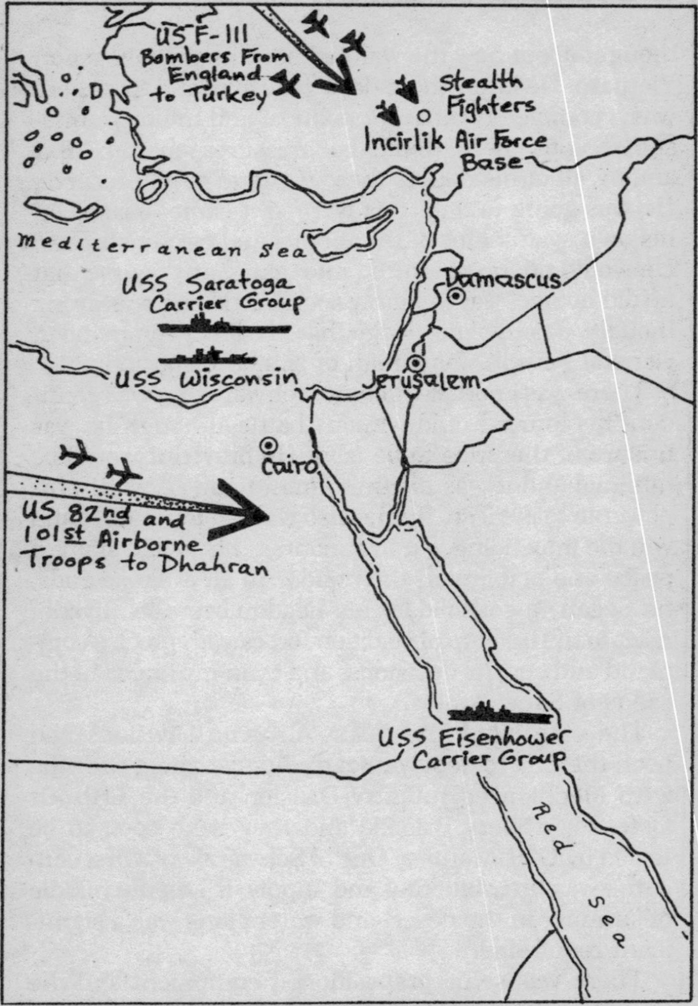

UNITED STATES ARMED RESPONSE TO IRAQI INVASION

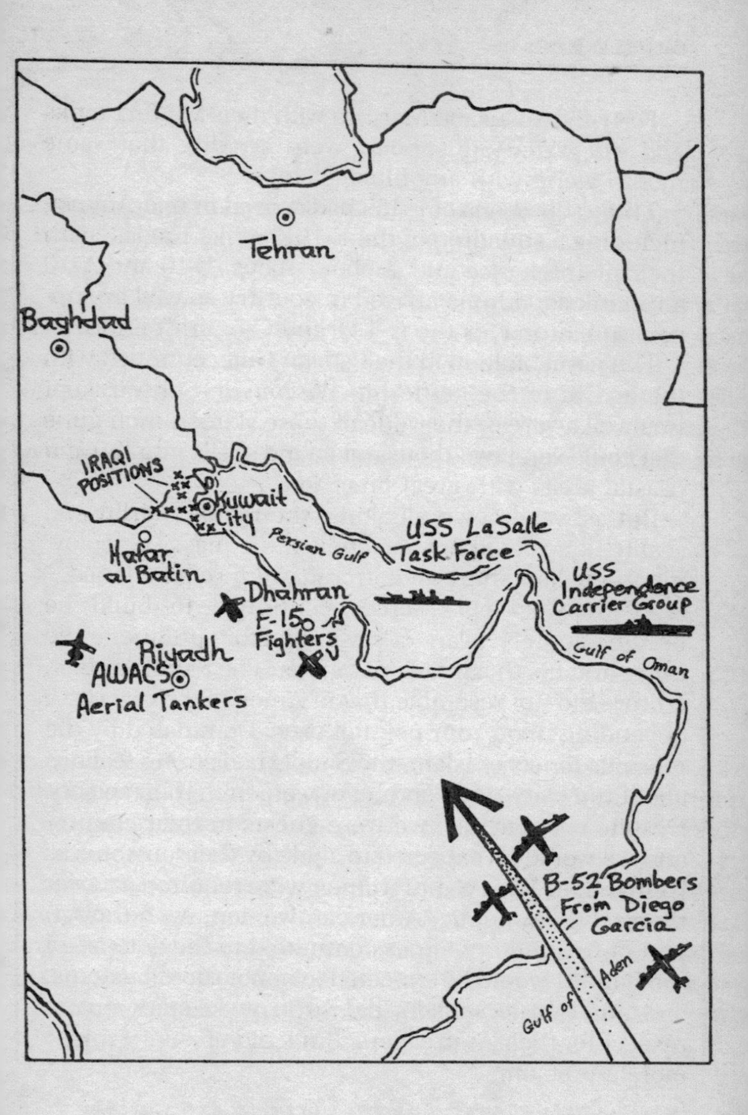
Tehran
Baghdad
IRAQI POSITIONS
Kuwait City
Hafar al Batin
Persian Gulf
USS LaSalle Task Force
USS Independence Carrier Group
Dhahran
F-15 Fighters
Riyadh
AWACS
Aerial Tankers
Gulf of Oman
B-52 Bombers From Diego Garcia
Gulf of Aden

Forty-five thousand Marines with their M-60A1 tanks and integrated air support were arriving that same month along with amphibious ships.

Three squadrons of F-15s had arrived in mid-August, including a squadron of the F-15E Strike Eagles, with their maintenance and support troops. F-16 and A-10 anti-tank squadrons arrived in country as well as support squadrons for the C-141 and C-5 airlift effort.

The naval buildup in the Persian Gulf continued with the arrival of the battleship *Wisconsin*. Schwarzkopf was well aware of the value of those sixteen-inch guns that could send two-thousand pound shells into Kuwaiti coastal areas with great precision.

But as was known all along, the military airlift and sealift alone could not accomplish the massive deployment; civilian ships and aircraft were requisitioned.

As Schwarzkopf's forces continued to build he became acutely aware of several factors unique to the times and his theater of operations. The Saudi Arabian culture did not resemble that of America, or vice versa depending upon your point of view. Dominated by the religious tenets of Islam, the Saudis had strong feelings about the place and conduct of women in their society. Despite our purpose, we were guests in their country and we would be expected to abide by their customs as well as their laws. Saudi women were required to cover themselves in public; American women, even though part of the military forces committed to the defense of the Saudis, would be expected to honor those customs, and our female soldiers did so in good spirit out of respect for their Arab sisters. But it did present a minor morale problem.

Alcohol was taboo and for the first time in American history, our soldiers, marines, and airmen prepared to fight a "dry" war! Even if one is a nondrinker, one can still appreciate the value of a cold beer in a hot desert, but such a treat did not exist. Instead, the consumption of cold soda reached record highs! Still, the absence of alcohol was another minor morale problem.

The practice of any religion except Islam was a sensitive issue. Force chaplains were ordered to keep a low profile and religious services were not to be widely publicized. But Schwarzkopf was determined his men would have an opportunity to pray and chaplains were always available to counsel the men and pray with them.

At one point, some troops were told to remove the American flags they had sewn on the upper arms of their uniforms. It was erroneously reported that it was a concession to Saudi sovereignty and the story generated some angry remarks back in the States. TV reports showed soldiers placing their flags defiantly inside their helmets, determined to wear them into battle. Later, it was reported that the removal of the flags had been ordered at a lower level by some commanders who were afraid that the Iraqis would concentrate on American targets. The matter generated some minor morale problems, but it was quickly solved and the shoulder flags reappeared. After all, American soldiers wore distinctively American battle helmets which were identifiable at greater ranges than the small flag.

There was a much more serious threat to good morale. Despite the record-pace deployment and rapid buildup, the troops experienced the age-old curse of soldiering: hurry up and wait. Despite being told that they might have to start fighting fresh off their airlift, the men and women of DESERT SHIELD found themselves shunted off to tent barracks and forward desert positions to hunker down and wait for developments. Nothing can take the edge off a fighting machine faster than boredom and lack of action. But Schwarzkopf had a remedy for that, and his orders were passed on to the field units. Train and rehearse. Move and dig in. Practice tactics and exercise command and control. Test and tax communications. Coordinate operations with other coalition forces and work out language difficulties. Conduct coordinated ground and air maneuvers. Once the 110-degree daytime heat gives way to the 95-degree night, conduct marches and practice defending against ambushes. Keep busy.

The delay did have advantages. Unlike Saddam's troops across the border, there were very few combat-tested men among the American units. Now would be the time to let the squad sergeants and young lieutenants make mistakes in deployment and command. Now would be the time to catch those mistakes and correct them.

Meanwhile, General Schwarzkopf settled into his headquarters. One of the first priorities was a red "hot line" to his boss, General Powell. They would confer every day. His personal quarters were adjacent to his war room and typically spartan: a picture of his family, and a cardboard box of audio tapes. An avid hunter and fisherman, he had brought the soothing sounds of wildlife along with his personal gear. There was the dynamic voice of Luciano Pavarotti, the homey twang of Willie Nelson, and the familiar sounds of Bob Dylan. The Bear's personal weapons were close at hand even when he managed to grab some much-needed sleep. His camouflaged-cover Bible was beside his bed and his stationary cross-country skiing machine he had brought with him from the States was nearby. Like so many of his age and build, he was constantly fighting the battle of the bulge.

Eighteen-hour days would be routine and even in his cramped personal quarters his mind would never be more than a blink away from his responsibilities.

He began his daily routine of adjusting his plan to developments and following the progress of his commanders in the field. The strategic responsibilities were his; the tactical execution was theirs. It was imperative that he and they were of one mind in the business of war, and early on Schwarzkopf made his intentions and method of command known. He was blessed with an experienced staff and outstanding component commanders, although it took a few sessions with The Bear to purge a few parochial suggestions and establish that the CINCUSCENTCOM was fully in charge of the activities of *all* the services.

Schwarzkopf was determined that his plan would provide for a massive coordinated attack by a truly unified command. He had vivid memories of Vietnam where at times it seemed like each service was fighting its own war. There would be no selfish priorities within his command. The Army, Navy, Air Force, and Marines—the forces of more than a score of nations—would fight with the tactics they knew best, but their efforts would be fully integrated within the battle plan of CINCUSCENTCOM. As Schwarzkopf put it, "I got a lot of guff." But when one goes jaw to jaw with The Bear, one better realize that the overall responsibilities are his and he's going to decide the issue. Arguments are welcome during the discussion phase (which may be involved or quite short depending upon the mindset of the CINC) but once the decision is made everybody will be expected to grab an oar and row the boat together. When Schwarzkopf's ground commanders arrived with their forces they had already studied the problems of their deployment to the field (as all good commanders would be expected to do) but they received something of a setback when Schwarzkopf told them otherwise. He would have them in one place for the DESERT SHIELD operation, then in another during the air war, and finally to their pre-attack positions as the date approached for the land battle.

When Air Force headquarters in Washington came up with an air attack plan—as they had been accustomed to do during the Vietnam fracas—Schwarzkopf's Air Force component commander, Lt. Gen. Horner, quickly informed his boss that if his people were going to execute the plan, they should have the major planning input. Schwarzkopf immediately added his thoughts and together, he and Horner came up with a plan as early as November that was exactly what they wanted and it would later prove to be completely executable and near-perfect tactically.

Schwarzkopf reviewed his overall plan constantly. It was uppermost in his mind as his forces continued to flood into Saudi Arabia and take their preliminary positions in the field. Even sleep failed to completely block it out. *It has to be the best*, he kept telling himself. And it would be. The Bear simply would not settle for anything else.

Fortunately, most of the routine headquarters load would be shouldered by Schwarzkopf's Deputy Commander, Army Lt. Gen. Calvin Waller. Waller, age 53, was confident USCENTCOM had a good battle plan, and with able commanders in the field there should be no need for any interference from the staff back at Riyadh. Their primary responsibility would be to insure that the battlefield commanders had what they needed to fight. He would also be instantly available should illness or any other unforeseen event incapacitate his boss. Of course, the entire staff knew that no germ or virus would dare attempt to penetrate The Bear's body and the CINC's four personal bodyguards would provide ample protection against any suicidal Iraqi—provided Schwarzkopf's battle expression didn't cut the guy in half first. Still, Waller would keep himself abreast of every facet of the campaign.

Marine Major Gen. Robert B. Johnson (53) would oversee headquarters operations as Schwarzkopf's Chief of Staff (COS). A native Scotsman, Johnson became a naturalized U. S. citizen at the age of 23, just five years after he immigrated with his parents. As the COS, he had the responsibility of overseeing the transfer of the headquarters and its 700 people and files and paperwork to Riyadh, a demanding task that he first undertook with a feeling of, "My God, how do we start doing this?" He had been on board for two months. Johnson and the other early arrivals in Riyadh had to settle for hotel rooms at first, actually running that phase of the operation from there until a Saudi government building could be made available, along with a schoolhouse. They quickly transformed their new quarters into a working headquarters and set up an armed-guard perimeter. They immediately felt a bit safer from any terrorist attack, which could have been a real threat when they were living on the economy, so to speak. Like the other key people, Johnson was on duty around the clock—as was his son, a junior officer in one of the marine infantry battalions out in the desert.

The Army component commander, Lt. Gen. John J. Yeosock, U. S. Army (53), an Army ROTC generated officer, would command more than 280,000 soldiers and not only be responsible for their overall tactical fighting performance but their welfare and care and all of the associated activities that went with a fighting army: targeting intelligence, tactical support operations, movement—in fact, the whole gamut of fueling, finding, and fighting. A cavalryman, Yeosock for a time commanded the renowned 3rd Cavalry, one of his predecessors being "Blood and Guts" George Patton. With all of his responsibilities, Yeosock likened his task to "running a corporation with one quarter or one third of a million people." Of course, his "corporation" had its "offices" spread across the sands of Saudi Arabia and was responsible for preparing the largest ground campaign since World War II.

The Air Force's component commander, Lt. Gen. Charles Horner, USAF (54), and a former fighter pilot who flew 48 Vietnam combat missions in F-105s and 70 more in F-4 Wild Weasels, tended to downplay himself as a man with no personality. For a fighter pilot that is just a plain contradiction of terms. He is a hard-working, hard-playing aerial tactician who found himself with a most important strategic task as the CINC's air boss. An old friend of General Yeosock, he shared a room with his army counterpart but they had little time to recount old times. Instead, Horner would concentrate on his command of an aerial force of over 2200 aircraft, some 400 of them allied contributions to the air war. One of his main tasks would be to coordinate USAF, USN, USMC, and coalition forces aerial sorties. He had been actually airborne, piloting an F-16 on a cross-country flight, when he had received orders to return to Shaw Air Force Base in South Carolina. That was on August 3. In short order, he assumed his USCENTCOM post, accompanied the CINC to Washington for a presidential briefing with Schwarzkopf, and headed for Saudi Arabia. He had entered the Air Force through the AFROTC program.

Vice Admiral Stanley Arthur, U. S. Navy (55), the USCENTCOM naval component commander, a combat-experienced naval aviator (500 Vietnam missions) and commander of the Seventh Fleet, was another excellent addition to the cadre of professional fighting men assigned as Schwarzkopf's senior commanders. The new boy on the block, Arthur assumed his USCENTCOM assignment in December, 1990, after the forces of DESERT SHIELD were in position, and his flag flew from the command halyard of the U.S.S. *Blue Ridge*, one of the Navy's sophisticated command and control ships. With a naval force of 120 U.S.N. ships, including six aircraft carriers, and 50 additional warships from eighteen different nations,

Arthur would direct a number of critical missions. While steaming his multinational force within the restricted sea room of the Persian Gulf, he would carry out embargo and blockade activities and conduct mine-sweeping of shipping lanes and operating waters, all the while providing naval attack aircraft for coordinated sorties over Iraq and Kuwait (over 3,000 would be flown) and rehearsing embarked marines for a possible amphibious assault. In addition, his forces would strike the much smaller Iraqi sea forces while protecting his fleet from the Iraqi air forces. In Arthur's cabin aboard the *Blue Ridge* was a hand-lettered sign, TMMP, translation: too many moving parts. And that was a reminder of one of the admiral's primary operating conditions: lots of ships and lots of airplanes all passing through the same airspace and seaspace on a 24-hour basis.

Rounding out the extraordinary cadre of USCENTCOM commanders, Marine Lt. Gen. Walter Boomer (52) brought to the team additional combat experience having also served two tours in Vietnam, one as a company commander and one as an adviser to South Vietnamese marines. He had learned well both the lessons and mistakes of that war. One of his pet peeves in that conflict had been the routine rotation of officers every six months to provide more of them with combat experience. That system "bordered on the immoral" by seasoning officers and then replacing them with new arrivals just when they were becoming combat effective. It was more than an idle observation. Boomer's Vietnamese troopers lost almost half their number in a 1972 battle that he considered a tragic example of poor planning. He was determined that his marines in Saudi

Arabia would not suffer from similar mismanagement. Soft-spoken but an aggressive field marine, Boomer would assign one of his major generals to sit in his seat at USCENTCOM headquarters so he could be out with his troops at a forward command post. As he settled down as marine commander, he had every intention of being just "a few vehicles behind one of my divisions" when the shooting started.

There is not an expert military leader in the world who can successfully fight a battle without the support of a capable logistics force. Schwarzkopf and Company had much more than that. Short of stature but long on ability, an intense logistician by the name of Maj. Gen. William "Gus" Pagonis, U. S. Army, 49, was key to the operation. He deployed to Saudi Arabia with the first headquarters contingent and for the first hectic days conducted his business out of the back of a four-door sedan. He and his advance team set a pace that overnight started bringing in everything from toilet tissue to tanks. Among his first priorities was the unglamorous task of insuring that ample latrines were built to insure the health of the troops. He set up a special mess tent to serve what most Americans would refer to as junk food. But for fighting forces at the front far away from home in an alien culture, hamburgers, chicken, and pizza are not just food but morale boosters. So what if the caloric content was on the wild side? The men and women of DESERT SHIELD would work it off in short order. And what they didn't work off, the midsummer desert sun would drain away through every pore in the soldiers' bodies. Water, perhaps the most immediate critical need, was provided in more than ample supply and one of the earliest images to come back to the States was that of troopers guzzling the precious fluid from one clear plastic bottle and carrying around another.

Schwarzkopf directed the activities of his staff and commanders with the firm-handed intent that his forces were not going to be caught off guard. The intentions of Saddam Hussein were unknown. Would he launch a preemptive attack? Was he preparing for a massive chemical or biological strike? Schwarzkopf knew that an Iraqi strike against the key logistics port of Dhahran or on his fighter-bomber airfields could be disastrous, therefore Schwarzkopf's first priority was to establish a ground and air defensive position and be prepared to hold it should the need arise while the remainder of his forces arrived in country. During the earliest days he knew that he was at risk. True, troops and fighting vehicles were arriving at a rapid pace but they had to take forward positions. The Saudis, of course, bolstered by the knowledge that their coalition allies were arriving to back them up, took the frontline positions while the allies maneuvered into position.

USAF tactical aircraft began flying combat air patrols within hours of arriving at airfields within Saudi. One of Schwarzkopf's early deterrents to an Iraqi assault were the carriers in the Persian Gulf. A formidable naval air force was always at the ready just a few miles off the coasts of Kuwait and Saudi Arabia, combat configured with their own munitions and support capabilities carried with them. Marines had arrived with their own integrated air support wings and more could be employed for amphibious operations on short order from the offshore amphibious assault ships.

Battlefield intelligence was of utmost importance and Schwarzkopf had several key sources. One of his primary needs-to-know was the disposition of Iraqi forces. Of equal importance was any indication that they were on the move. His staff began scrutinizing satellite photos and prepared daily disposition reports of Saddam's ground forces. USAF E-3 airborne surveillance aircraft (AWACS—Airborne Warning and Surveillance System) searched the skies for any enemy aircraft movement or hostile intent. Some frontline intelligence could be gathered by forces in the field and Schwarzkopf made plans for the use of Special Forces once they became available. Communications were monitored, both from tactical units deployed in Kuwait and Iraq and from command and control headquarters in Baghdad. At the CINC's headquarters a picture was being built and when it was complete, Schwarzkopf could adjust his own strategic plan. He had a basic idea of what he would do if forced to fight defensively, but he needed to fine-tune his thinking should the economic sanctions approved by the United Nations fail to force Saddam to withdraw his forces from Kuwait. He had one goal in mind either way. If he was forced to fight it was going to be all-out with no punches pulled. He was confident of victory under any circumstances but the longer he had to build his force, the more opportunity he would have to keep down casualties. And that remained a prime concern. There was already too much talk of body bags back in the States, mostly by the media and dissenting political and public figures. America was not completely united about the possibility of using military force but the majority that did support the deployment of USCENTCOM was growing daily. That was encouraging. Schwarzkopf wanted the great mass of American people behind him and his troops. There would certainly be a few who wore Vietnam blinders and had neither the perception nor the inclination to realize the great differences in the Persian Gulf situation. He could live with that.

There was a small segment of the U.S. population that was vocally opposing any form of military action, whatever the provocation. *No blood for oil!* was an early cry. But the impending conflict was not primarily about oil, although there was obviously a peripheral relationship. The situation was one of aggression and a threat to the stability of the region. Iraq had brutally invaded, occupied, and intended to annex the independent kingdom of Kuwait. Saddam Hussein had demonstrated a personality that was chillingly reminiscent of another power-hungry despot, Adolph Hitler. Could Kuwait be a modern-day Czechoslovakia? Iraqi forces were massed along the northern Saudi border and King Fahd had asked for help. Practically all of the nations of the world had condemned the Iraqi invasion of Kuwait, a solidarity never before witnessed in such a grave matter, and a number were responding to Fahd's plea.

Schwarzkopf was aware of the political climate back home, but overall it was favorable and he was encouraged by the support the man in the street—whether in New York, Toronto, London, or Paris—was giving him and his troops. Poll after poll was revealing that people in the coalition countries felt their governments were on the right track. Even the dependents and parents of the service personnel were expressing overwhelming support. Besides, for the moment CINCUSCENTCOM had much more important things on his mind.

If he were to fight a defensive war, he would be ready. He had already decided that he would lure the Iraqi forces into the desert as far as he could, then pound them with his superior airpower and artillery. Finally, he would attack Saddam with his armor. As he put it to a group of reporters while flying out to visit some of his Special Forces, "I'd engulf him and police him up. It's that simple." But although Saddam was making threatening noises, his troops were digging in for a defensive posture themselves. The threat of an Iraqi incursion into Saudi Arabia was lessening. The media were unwittingly playing a role in convincing Saddam that such a move would be very unwise. Published reports by eager reporters anxious to provide more information than their competition to the public indicated an awesome influx of coalition forces into Saudi Arabia. These reports were somewhat misleading, for although forces were indeed pouring in, they were not yet in position to fight and some acclimation would be necessary for maximum fighting ability. American troops simply were not familiar with the particular desert terrain of Saudi Arabia although many of them had received desert training. Schwarzkopf knew that Saddam Hussein was also well aware of published media material. After all, TV networks were broadcasting every tiny bit of information they could get their hands on and satellite dish reception in Baghdad was excellent.

Schwarzkopf was determined to stay ahead of events and there was an opportunity, even during the August buildup, to take further advantage of the openness of American society. Only this time it would be a deliberate deception designed to mislead Saddam in his analysis of coalition battle plans.

Coalition naval forces were in complete charge of the gulf and U. S. Marines were ready. If hostilities should break out, they could very well be called upon to assault Iraqi forces in Kuwait with a combined amphibious and airborne attack. It was always beneficial to run a rehearsal before the actual tactic and Schwarzkopf approved a series of naval and marine maneuvers that culminated in operation Imminent Thunder, a full-scale landing on and across the beaches of Saudi Arabia very close to the actual seacoast of Kuwait.

There were hundreds of media representatives covering the coalition forces, perhaps as many as ten times the number that performed the same function in Vietnam. There was no way that Schwarzkopf could keep his military movements classified (i.e. secure from detection by Saddam) if the media reps were allowed to swarm unrestrained throughout the theater of operations. At the same time, the CINC was well aware of his responsibilities with respect to the public's right-to-know within the provisions of the U. S. Constitution. What some of us tend to forget is that the precious piece of paper also contains a restriction with respect to the rights of its citizens, that of giving aid and comfort to the enemy (Article III, Section 3). If done so voluntarily, in grave matters, giving aid and comfort constitutes treason. But in wartime even an inadvertent or careless release of certain military information could involve the giving of aid and comfort to the enemy and would most certainly result in a compromised tactical operation and/or increased casualties. Therefore, it was necessary for USCENTCOM to make a judgment as to what military information fell within the need-to-know category and what would aid the enemy. Certainly, matters that carried a security classification could not be released for public consumption.

To that end, reporting pools were formed, each consisting of a small number of media representatives. The pools would be allowed to cover certain military actions and forces with the understanding that the coverage would then be made available to all of the media personnel present in-country. Those not assigned to pools would have access to several military briefings held daily in Riyadh, one by USCENTCOM and two others by British and Saudi commands. In addition, unaired background briefings were provided to the media. Otherwise, travel within the theater of operations by media reps was prohibited.

Brigadier General Richard Neal, USMC, Schwarzkopf's Deputy Director of Operations, was assigned as the principal headquarters briefer for the progress of DESERT SHIELD/STORM operations. A lean, articulate New Englander, Neal was consistent in his approach to the briefings. He would read prepared material about the previous day's activities and open the briefing to questions. He refused to speculate on operations, confining his remarks to facts of which he had

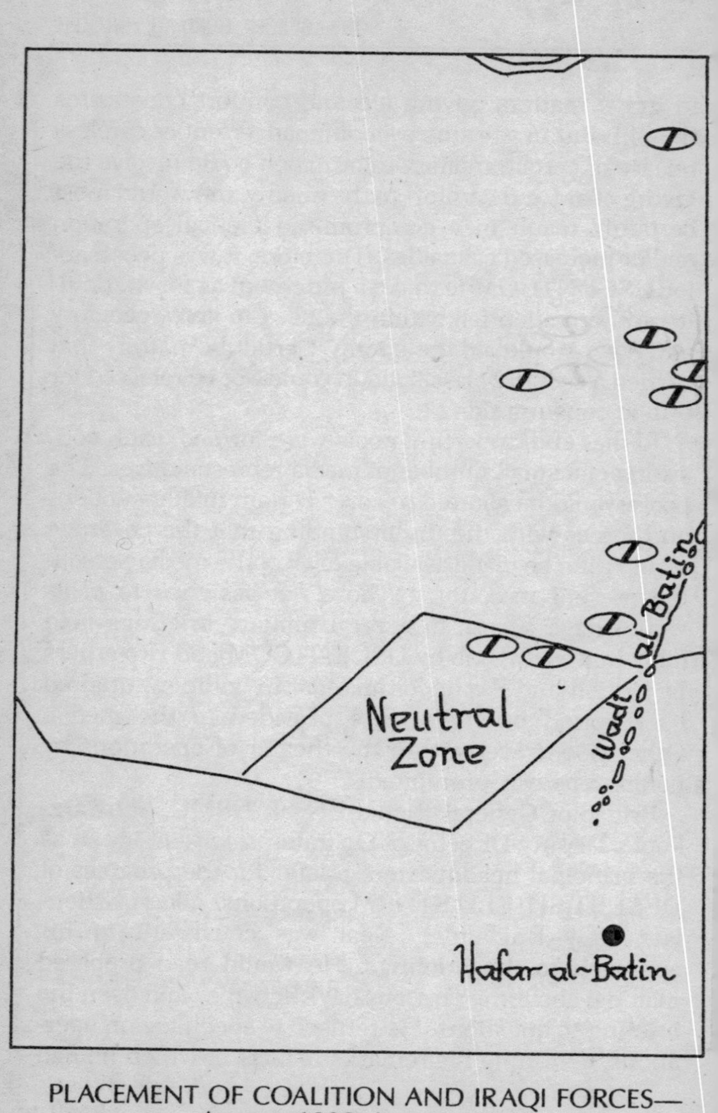

PLACEMENT OF COALITION AND IRAQI FORCES—
August, 1990–January, 1991

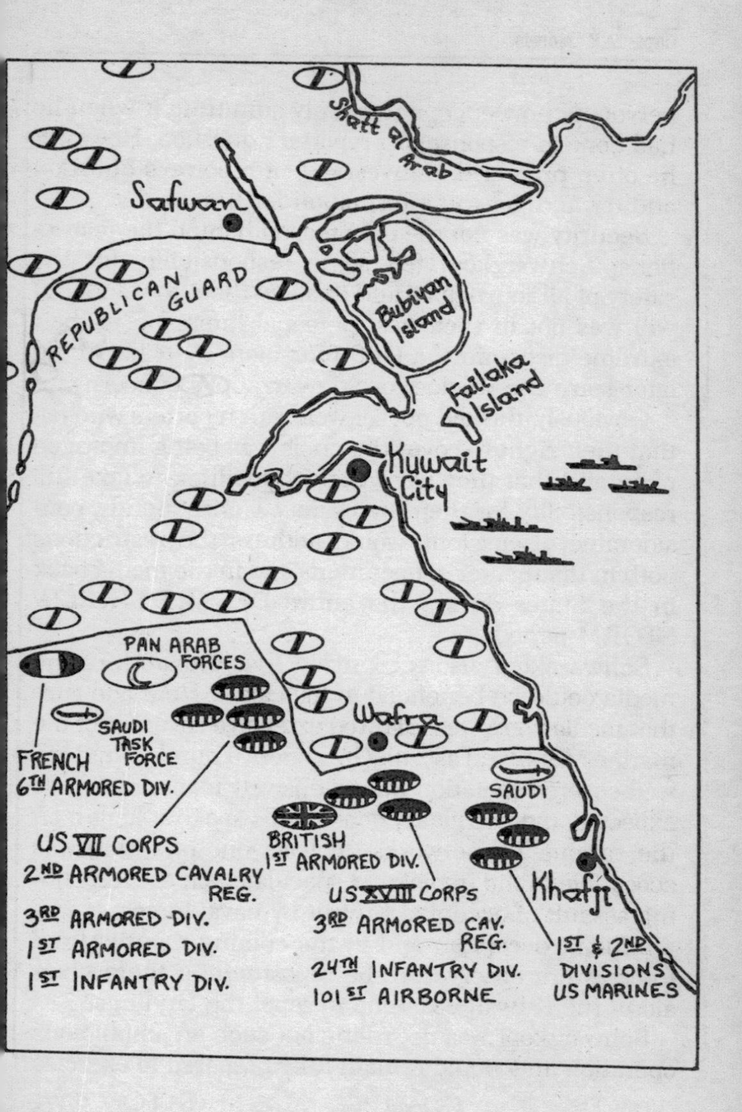
Shatt al Arab
Safwan
REPUBLICAN GUARD
Bubiyan Island
Failaka Island
Kuwait City
PAN ARAB FORCES
Wafra
SAUDI TASK FORCE
FRENCH 6TH ARMORED DIV.
SAUDI
BRITISH 1ST ARMORED DIV.
Khafji
US VII CORPS
2ND ARMORED CAVALRY REG.
3RD ARMORED DIV.
1ST ARMORED DIV.
1ST INFANTRY DIV.
US XVIII CORPS
3RD ARMORED CAV. REG.
24TH INFANTRY DIV.
101ST AIRBORNE
1ST & 2ND DIVISIONS US MARINES

personal knowledge, and openly admitting it when he had none in response to a reporter's question. However, he often promised to investigate a reporter's question and try and get the information for him or her.

Security was not the only reason behind the restrictions; Schwarzkopf felt some responsibility for the safety of all journalists and to allow them to wander at will was not in their best personal interests (to their extreme discomfort, a CBS four-man TV team would later learn the wisdom of Schwarzkopf's concern).

Obviously, this did not set well with reporters who felt that their right to cover the crisis was being impinged upon and that they were perfectly willing to take full responsibility for their own safety. Consequently, considerable discussions would address the restrictions both in the theater of operations and in the media back in the States during the entire DESERT SHIELD/ STORM period.

Schwarzkopf also recognized that the work of the media could be beneficial to his cause. He made sure that media pools were allowed extensive coverage of the marines' rehearsal assault (Imminent Thunder) and the well-covered operation was extensively reported. As was expected, media military consultants spoke at length on the various TV networks (at least one of which was accessible to the Iraqis) and speculated on the impending assault of occupied Kuwait by naval forces. It was a brilliant deception and in the coming months Iraqi forces were observed to be repositioning themselves along the Kuwait coastline to repel the landings.

Schwarzkopf was not ruling out such an amphibious operation and would remain fully prepared to exercise that option if needed but his main purpose was to mislead the Iraqis and draw more of their forces eastward toward the sea. As for the actual attack, the CINC had another overall strategy in mind and the first step toward insuring the success of that strategy had been taken. A significant number of Iraqi troops—some nine divisions—would be positioned far away from the main assault.

CHAPTER 3

AS THE DESERT days began to shorten with the coming of September, General Schwarzkopf began to breathe a little easier. His force buildup was continuing and still had quite a way to go, but he was no longer in the precarious position of the earliest days of DESERT SHIELD. Regrettably, during the early days of the deployment, the operation had claimed its first casualty. Air Force Sergeant John Campisi had been struck by a truck as he went about his nighttime duties on one of the Saudi airfields.

Had Saddam decided to strike across the Kuwait border into Saudi Arabia in the hectic days just after the first U.S. troops arrived in August, it could have been a very rough time. The political ramifications back in the States not only might have weakened support for

USCENTCOM's task, but even forced further negotiations with Saddam Hussein. The whole operation could have been halted. But that had not happened and Schwarzkopf, whose post-service ambitions included becoming a champion skeet shooter and a top salmon fisher, continued his efforts in the critical task of consolidating his forces. He and his staff also monitored political developments with daily attention. Surely there was still the possibility of a peaceful settlement to the crisis.

Schwarzkopf used every opportunity to get out into the field and talk to his people. It was a morale booster for them and therapy for him. The visits always lifted his spirits and made him even more confident of the fighting ability and professionalism of his troops. True, very few of them had been in combat, but he was convinced that they had received the best training available, and their equipment was among the most sophisticated in the world.

All of his forces were volunteers. That was the nature of the military in the nineties. They had undertaken their responsibilities on their own; not a one had been forced to serve. Several prominent people back in the States were making noises that if war came it was going to be fought by the poor and underprivileged. Nothing could have been further from the truth.

Almost without exception, the American fighting man or woman had at least a high school education. Once in the service, they had received extensive training and further education. Experience with high technology reached all the way across the military spectrum from infantryman to tank crew and, while a number of troopers may have come from low-income environments, they were to a man motivated and knew that the service provided an opportunity to better themselves. They would prepare for return to civilian life with skills and a mental attitude that would enable most of them to continue an aggressive pursuit of the American dream. It is doubtful that they looked upon military service as an escape; more like an opportunity. And for segments of the press or public to belittle them by inferring that the United States was taking advantage of its poor by using them to "fight a war for oil" was not only unfair, it was insulting.

A great number had educations above the high school level. Certainly a significant segment, primarily the officer corps, had college educations and advanced degrees, and to overlook that and infer that battle would be confined to enlisted troops showed a gross ignorance of the ways that wars are fought.

Schwarzkopf was very impressed with his young officers and one haunting thought was the knowledge that when it came to combat casualties, the second and first lieutenant platoon leaders would suffer the highest casualty ratio. That was not just a statistic to him; as a junior officer, he had ridden that train across the bloody tracks of Vietnam on two separate occasions, earning his two Purple Hearts.

The coalition air forces would fight the first battles, and the proportion of highly educated and extensively trained service personnel that would be killed and wounded was staggering compared to other losses. One had only to review the statistics on deaths and captures in previous air wars to see that the sons of upper-class and middle-class America carried more than a proportionate share of casualties in an air war. No one was screaming about that. A review of the 1962–1975 guest book at the Hanoi Hilton (North Vietnamese prison camp at Hanoi) was a quick and ready proof of that claim and even a hasty glance at the video of returning POWs from southeast Asia more than a decade back should have been a reminder that *everyone* in combat shares a common risk.

As for the equally asinine claim that the government would be deliberately using a disproportionate number of blacks to fight the impending war, the fact that all were volunteers should be a source of great pride for America, not concern. The black soldier, sailor, and airman all saw in the services an opportunity to overcome adversity, learn self-discipline and most probably a trade, and better themselves. And the military had been a leader in providing that opportunity to all Americans. An improvement in the quality of life was a universal goal. And if the path included the most serious of obligations, it provided for an even more worthy accomplishment.

As Schwarzkopf made his visits, he was particularly careful to avoid disrupting his troops' routine, for they had much to do to prepare for combat. He knew that any time a four-star was scheduled to visit a unit there would be a flurry of activity spent shining brass and policing the area. So, instead, he tended to drop in on units with a minimum of fuss. He was there to talk and listen to his troops, not to inspect them. And if his visits provided a bit of inspiration (and they did!), that was great.

A number of pictures of The Bear surrounded by his men and women reached the American public through the various media, and we all enjoyed the human side of a special warrior. Of course, Schwarzkopf also picked up a few remarks on these visits that indicated problems needing his attention. Boots were one example. The standard issue footwear was an old design, intended for the hot humid climate of southeast Asia. In the desert, sands tended to slip through the ventilation holes, and the thick leather didn't allow the feet to "breathe" as much as they needed to in hot, dry Saudi Arabia. The old axiom may state that an army travels on its stomach, but in reality it also travels on its feet—despite the highly mobile capability of mechanized and cavalry units. The Bear made the acquisition of new boots for his men a personal priority, and he saw that a number of the soldiers had acquired the Saudi boot, a lighter covering designed for desert use. The CINC made a personal evaluation and soon ordered his logistics people to start procuring a more suitable boot.

With the possible exception of the old-styled boot, the American soldier and marine were well equipped for their role. A bullet-resistant Kevlar helmet protected the head with a design that was somewhat reminiscent of the German "pot." The new headgear provided more protection and was lighter than the old style. One piece, it lacked the versatility of the old two-piece style, which had an inner plastic shell and an outer shell that could also be used for washing and shaving water. Over his desert camouflage outer garment, the trooper wore a Kevlar flak jacket and a load-bearing harness that carried ammo, a combat knife, first aid supplies, and a

9-mm Beretta pistol. His rucksack held two ponchos, a poncho liner, rations, underwear and personal hygiene items. Leg and hip pockets on his outer garment provided additional personal stowage space. An entrenching tool was strapped to the side of the rucksack and a pair of canteens hung from his combat belt. Strapped to the bottom of the pack was his bedroll, and he carried the standard issue M-16 A2 automatic rifle. There were slight clothing and equipment variations between the Army and Marine troops, but basically the individual fighting man was largely self-contained except for the reprovision of expendable supplies. Finally, there were two new items: a protective mask to filter out chemical and biological agents, and a protective suit (stowed in the rucksack).

In his headquarters, Schwarzkopf followed daily reports on the progress of his force buildup and kept a weather eye on world developments. He would not give up his hope that war could be avoided, but there was increased evidence that Saddam was not thinking of backing down.

Saddam had declared that the embargo was an act of war and boasted to the world that if the Americans (he shied away from ever referring to the coalition) dared to fight him, there would be "thousands of Americans in sad coffins." Intelligence reports indicated that in excess of 160,000 Iraqi troops could be in Kuwait.

Saddam abruptly ended his war with Iran, permitting the release of thousands of troops for redeployment to southern Iraq and Kuwait.

President Bush authorized a call-up of selected reservists and naval forces were authorized to use a stop-and-search tactic in the enforcement of the sea embargo. More and more statements were being made that Israel would "inevitably" be drawn into any conflict and Schwarzkopf knew that such a development would place a great strain on the coalition. There was a distinct possibility that certain coalition Arabs (the Syrians, for example) would refuse to fight *with* Israel against Iraqi Arabs. Such an event would effect more than just a reduction of numerical forces; it would shatter the historic accomplishment of Westerners and Arabs fighting side by side.

Saddam was steadfast in his intent to hold thousands of foreign nationals, primarily Americans and Britons, as "guests" of the state. In fact, Saddam said on August 17, they would be housed in strategic military locations to deter American attacks on those installations. Obviously, this was of great concern to Schwarzkopf. If war came, he would have to strike some if not all of those targets. And just as obviously, although Saddam refused to use the word, the detained people were hostages, not "guests." Saddam was being his usual satanic self.

Jordan was drifting away from its sympathetic relationship with the U. N. resolutions and actions, and openly siding with Iraq. There was always a possibility that King Hussein would come to the military aid of madman Hussein. Already, some embargoed items were slipping across the Jordanian-Iraqi border.

There was every indication that the Iraqis had stockpiles of chemical and biological weapons, and it was necessary for coalition forces to conduct some of their training exercises in full chemical/biological gear. Obviously, the cumbersome suits and limited-vision masks affected the mobility and awareness of the troops and drilling with them in the hot sun was often an ordeal in itself. But they did provide protection against all of the known agents, and the men carried syringes of antidotes for self-injection in the event of exposure.

On the plus side, Schwarzkopf felt confident that he had enough forces in-country or at sea in the gulf to hold off an Iraqi thrust. The first M1A1 Abrams tanks had arrived. Troop morale was high and during his field trips he was constantly hearing such remarks as "Let's go get 'em!" "We're gonna kick butt!" and "All the way!"

There were a few laughs.

There was a report that a group of British soldiers had called the Hilton in Baghdad and asked for the reservation clerk. Once she got on the line, they asked for room reservations for the end of February—and they wanted the best rooms—"with a view!" The flustered clerk hung up without replying. Their call was uncannily prophetic, although Schwarzkopf had no orders or intentions to insure their reservations.

Baghdad Betty, a young English-speaking female broadcasting on Baghdad Radio in the tradition of Tokyo Rose, cranked up her spiel—but quickly lost credibility when she warned American servicemen that their wives and sweethearts were being unfaithful and sleeping with celebrities including "Bart Simpson."

As September wore on, there were also some interesting developments. On the 28th, Saddam reversed himself and indicated that he would let the women and children hostages leave Iraq—and even the men—if the U.S. would pledge not to attack Iraq. And he ordered his ships to submit to boarding and searches.

Japan followed the lead of Saudi Arabia and pledged funds to offset the gigantic costs of the operation—offering one billion dollars, not quite what one would expect from one of the world's leading economies. Prime Minister Kaifu's reply to criticism was "We did our best." Later, they would do more.

Then, in another turnaround, Saddam sent additional forces into and around Kuwait for a total of some 265,000 facing the coalition. Saddam declared that if war came, one of his first targets in addition to Saudi Arabia would be Israel. That caused new consideration of an old question: Would Israel retaliate? If she did, what effect would it have on the coalition? The likeliest answers to those questions were unsettling, indeed.

A thousand questions went to bed with Schwarzkopf whenever he slept. He would have preferred to dream of walking through the brush with his black Labrador retriever, Bear (what else!), or popping a few clay pigeons with his favorite shotgun, or breathing the cold crisp air of Alaska while pulling record salmon from frigid streams. It was the dilemma of every battleground commander in his brief moments away from immediate responsibilities. One wanted to relax and recall the good things in life, but they were in another world and in another time. However, they were never completely lost, and they were the comfort of not only Schwarzkopf but of every soldier involved in DESERT SHIELD/STORM. The general tried to phone his wife whenever the opportunity presented itself, but it was seldom more than once or twice a week. He delighted in hearing family news and always inquired about the health and welfare of everyone back home. He had never made a practice of taking his work home from the "office" when they were together and, even now, he continued that practice by not talking about his military duties.

Over 30,000 American reservists were called to active duty. Over 8,000 had been summoned in August and current law allowed for the activation of a total of 200,000 with some time restrictions (90 days without approval of Congress), and a number of these would be joining regular forces when sufficiently trained. As if to underscore the hazards of just getting to the war, a giant C-5 *Galaxy* airlifter crashed in West Germany and nine reserve crewmen were killed.

Despite diplomatic maneuvering and Saddam's up-and-down attitude toward finding some political solution to the crisis, the allies stood firm on their commitment to carry out the resolutions of the United Nations, the most demanding of these requiring Iraq's withdrawal from Kuwait. As September slipped into its third week, President Bush reemphasized his determination: "Iraq will not be permitted to annex Kuwait. That's not a threat, not a boast. That's just the way it's going to be."

The Iranian religious leader, Ayatollah Ali Khamenei, issued his own statement, "Anyone who fights America's aggression has engaged in a holy war in the cause of Allah, and anyone who is killed in that path is a martyr."

General Schwarzkopf had a new question to ponder. Now that Iraq and Iran had officially terminated their war, were the Iranian leaders preparing their people for an alliance with their recent enemy? If anyone hated the Great Satan more than Iraq, it had to be Iran. Weakened by eight years of unsuccessful war against the Iraqis, the Persians could not add significantly to the military forces arrayed against the coalition. The real purpose of the Ayatollah's statement seemed to be a call to arms for certain fanatical elements of Islam. At the very least, it seemed there was a renewed call for worldwide terrorism.

But Schwarzkopf received some encouraging news as well. Several stateside polls showed that an overwhelming majority of his fellow citizens now approved of USCENTCOM's mission—75 percent, in fact. Enthusiastic displays of patriotism were spreading across America, and the sale and display of flags was record-setting. The American people were already determined to let the troops in Saudi Arabia and the gulf know the country's support was behind them. Yellow ribbons began to appear in great numbers.

And the French were sending more troops (to bring their total to 5,000), as were the British (new total: 9,000). The numbers were welcome and they were indicative of the strengthening and determination of the coalition. Even Syria agreed to dispatch 300 tanks and a total of 18,000 soldiers.

Schwarzkopf, the diplomat, examined his command and control chain and his order of battle. If war came, he would insure that Arab troops fought under their own commanders in concert with other coalition forces and would not be employed in any way that would be detrimental to their national pride. A particular effort would be to insure that the Kuwaitis entered Kuwait City first if tactically possible. The Saudis, proud defenders of their land, would play a major and leading role. And the other Arabs would be teamed up with Westerners with whom they felt comfortable. It was not a simple task to balance these needs, for by this time, 26 countries were lending military support.

Notably absent were the Soviets, although they had joined the world community in condemning Iraq. The U.S.S.R. had decided that the only way they would send troops was if the operation was placed under a U. N. commander and became a U. N. force. But it was a bit late for that. To meet this criterion, an entirely new headquarters and staff would have to be designated and a completely new chain of command formed. The United States was not about to let anyone else decide when and how their forces would be taken into this particular battle. The designation of CINCUSCENTCOM as the overall commander was perfectly in accordance with U. N. provisions for bilateral and multilateral defense arrangements. No one said it, but a lot of people thought it: The Russians weren't needed. Their presence on the battlefield would cause all sorts of problems, although in truth it would have been an historic achievement, American and Soviet soldiers shoulder-to-shoulder, charging across the battlefield. Besides, there were indications that the Soviets were trying to play both sides of the conflict. They were not universally in agreement with the coalition actions and perceived intentions, and they didn't want to completely alienate Iraq, their biggest military hardware customer. If war did come, they knew as well as everyone else that the coalition would win, and part of that winning would be the destruction of much, if not all, of Iraq's military hardware. After the peace, there would be a new market for the products of the troubled Soviet economy.

As the month of September drew to a close, coalition force strength began to approach that of the Iraqi numbers arrayed in and around Kuwait. Schwarzkopf had approximately 240,000 coalition troops (with the number increasing daily) and 700 aircraft, as opposed to 360,000 Iraqi troops and an estimated 500 aircraft. But Schwarzkopf was still behind in tanks and armor. That discrepancy would be corrected as the desert cooled in the fall months and environmental conditions for battle improved.

October found Schwarzkopf again reviewing his plan and options. There were times when he had doubts. Suppose the plan didn't work and casualties were unacceptably high? That was the nightmare of every planner. There were other times when he felt confident that he and his staff had considered all possibilities and all that was left to do was react to any changes in Iraqi strength and disposition. Overall, however, he knew that he had the personal intelligence and the professional background required for the job. And his personal goal remained firm. He would bring the military crisis to a successful end and he would do it with a minimum loss of life, including that of his enemy.

There comes a time when the commander must stop trying to second-guess his own plan and say to himself, *That's it. It's a go.* It can be a very difficult decision, but that is what leadership is about. That is why a commander must assert himself, sometimes with blunt authority. A military leader isn't in a personality contest—but invariably a great military leader is respected and admired, although it may be after the fact.

Some compared Schwarzkopf to Lawrence of Arabia. Like his famous predecessor, the general had forged an unlikely alliance among the Arabs, and in years past when he was on a military liaison trip to the Middle East, he had even put on Arab dress at the insistence of his Kuwaiti hosts. By all accounts, he had enjoyed the sensation of the flowing, loose-fitting garment.

By mid-month, Saddam had dispatched more troops to southern Iraq and Kuwait, and coalition forces could now be facing almost half a million men across the northern Saudi border. The average Iraqi field soldier was not the equal of his coalition counterpart from the standpoint of either training or equipment. It was assumed that he was devoted to Saddam and was more than willing to die in battle, bolstered by his religious belief. The men of Saddam's special force, the Republican Guard, were considered even more loyal and much better trained. The media almost always referred to them as the "elite Republican Guard," but experienced coalition personnel considered that to be relative with respect to Iraqi forces overall. The Guards were certainly the most loyal and the best equipped.

Schwarzkopf was well acquainted with the Iraqi weapons and their capabilities. Of particular concern, of course, was Saddam's resurrection of chemical and biological weapons from the graves assigned them by the Geneva Conventions. Missiles and munitions capable of delivering chemical and biological agents were on hand to employ the deadly agents outlawed by civilized nations since World War I, but brandished now without hesitation by a madman.

But Schwarzkopf knew that there was an up-side to any planned use of chemical and/or biological warfare by Iraqi forces. To begin with, as horrible as the weapons were, chemical and biological weapons are the only ones that can be defeated by alert, protected personnel. Any other weapon (bullet, bayonet, bomb, shell, etc.) will cause a casualty on contact. The dry desert environment of Saudi Arabia, southern Iraq, and Kuwait, was not conducive to the use of chemical and biological weapons. Saddam had a limited supply of such weapons and they could be expected to arrive in relatively small doses via projectile, bomb, or warhead. Allied forces had overwhelmingly retaliatory capabilities with conventional arms and the average Iraqi soldier had little or no self-protection if he used chemical and biological agents. Finally, once the war started, Schwarzkopf intended to close on the enemy very rapidly and in close combat: in such circumstances, chemical and biological weapons can affect friend as well as foe.

Saddam's military hardware was almost exclusively Soviet. The popular AK-47 infantryman's weapon was an excellent weapon but surely outmatched by the American M-16 A2.

His artillery included the 155-mm G-5 howitzer with a formidable range of over 20 miles. Among Iraqi mobile systems, the Brazilian-designed Astros II SSM (surface-to-surface missile) rocket system had a range of 30 miles. The Soviet FROG-7 SSM system was old but had an impressive range of 55 miles and was known to be capable of employing chemical warheads. The "big gun" Scud SSM could be launched from either fixed pads or mobile units and was a real terrorist-type threat, with its range measured in hundreds of miles. But it was too inaccurate to present a viable military threat. The Iraqis also had a limited number of the Chinese Silkworm SSMs.

The MILAN, a portable antitank weapon of French and Belgian design, featured a heat-comparison night sight. For antiaircraft defense in the field as well as in the cities, there was the four-barreled ZSU-13-4 antiaircraft gun. Self-propelled and with a radar-sighting capability, it was a formidable around-the-clock weapon.

Soviet BTR-60s, akin to our own armored personnel carriers, were standard infantry fighting vehicles.

The heavy armor consisted of the battle-tested Soviet T-72 main battle tank with its powerful 125-mm gun, and the older and less sophisticated T-62 with its 115-mm gun and less accurate sighting mechanisms, which was the mainstay of Iraqi tank forces.

Saddam possessed an impressive array of aircraft, the vicious late-model MiG-29 Fulcrum fighter-bomber being the cream of the crop. It could carry guns, air-to-air missiles (AAMs), rockets, bombs, or combinations of these weapons.

The Mirage F-1 could carry laser-guided weapons as well as AAMs and could deliver the feared French Exocet air-to-surface missile (ASM).

The Soviet Su-24 Fencer ground-attack aircraft could attack not only in daylight but at night and during bad weather, due to its sophisticated detection and aiming electronic systems.

Airlift was the sole mission of the large Il-76 Candid

airlifter, although it could be configured to act as an early-warning radar platform.

Finally there was the impressive Soviet Mi-24 Hind attack helicopter (personnel and antitank) and the French SA-342 Gazelle, also an attack helicopter.

Saddam's naval forces were minimal, consisting primarily of a few lightly armed coastal-patrol boats.

Schwarzkopf had every confidence that coalition weapons, particularly those of American, British, or French design, were superior in every respect. His problem would be to have the weapons in sufficient numbers to attack a dug-in defensive army with armor and air support. With an engineering background, he was as familiar with the high technology of current weaponry as any commander that ever took the field.

Most of the coalition's command and control problems had been solved, but there was some concern about language difficulties. Rapid and precise communications are a must on the battlefield where there are enough equipment, operator, atmospheric, and environmental glitches without having two different languages on opposite ends of a communications link.

There were also concerns about the fact that Saudi military commanders had been appointed for their loyalty to Fahd rather than any demonstrated military prowess. Most were relatives of Saudi royalty.

The other coalition Arabs were placed under the direct command of fellow Muslims: The Kuwaitis under the Saudi chief of staff, General Khalid bin Sultan; and the Syrians under the commander of the Egyptian Corps. Along with consolidating religious and ethnic groups, such assignments also minimized some potential tactical control and logistic problems. The Saudis used American hardware and tactics while the Syrians, for example, employed Soviet equipment and their tactics reflected Soviet tactical doctrine.

On October 23, Schwarzkopf's morale received a big boost from the visit of his immediate boss, General Colin Powell. Undoubtedly, General Powell received an equal injection of adrenaline from the opportunity to get away from the endless halls of the Pentagon and visit the forces in the field. And the troopers were excited to hear some comments from the horse's mouth about the status of things back in the "puzzle palace."

The two top military men briefed each other on matters within their own spheres of responsibility, and Schwarzkopf was able to present the latest revisions of his plan to his boss. Powell heartily approved and recognized once more the brilliance of his commander. Each felt comfortable with the other and each respected the judgment of the other. One journalist likened the pair to a modern Eisenhower and Patton and one could see a basis for that comparison. They were indeed a team; each had contributed to the excellence of the plan and the manner in which it was being carried out.

During November, as in all the months preceding the execution of DESERT STORM, there was always the danger of an inadvertent incident setting off a battle that could escalate into a full-scale and premature engagement. Such a forced entry into war could knock all plans to the winds and had to be a constant concern of Schwarzkopf, his staff, and his commanders in the desert. Who knew when some crazed Iraqi pilot might take it upon himself to jump in his aircraft and head over Saudi Arabia or some fanatical field unit decide to take on coalition forces close to the border?

As insurance against the former case, allied fighters were always airborne or on alert to intercept any unexpected aerial threat; and front-line intelligence units were alerted to look for any signs of the latter.

Many of the troops who had arrived in early August had been in the field for over three months, and one can only sit on the edge of a sword for so long without falling off. Still, there was a general consensus that as boring and as tiresome as waiting can be, the alternative, battle, was less desirable. There were cries of "Let's get on with it!" but that was mostly high-spirited bravado masking the real dread of a war that could come at any minute. It is a real credit to the coalition forces that amid all of the waiting and uncertainty and confusion, they remained alert for the entire period DESERT SHIELD was in operation. That is truly professional performance.

One has to take a moment and reflect on what some commentators have referred to as the "new American military." It would appear that the U. S. has learned some very valuable lessons from Vietnam. Certainly, it learned heartbreak. The tragic way in which that military campaign was politically controlled not only robbed them of the effective support of the South Vietnamese but stamped the military with a stigma that is just now being erased. But they still carry the scars. General Schwarzkopf, when asked on a television interview in March, 1991, on whether the great success of his campaign in the Middle East had erased his scars—and he has both physical and mental souvenirs of those days—responded, "Oh, I don't think it'll heal the scars of Vietnam. Those scars are there. It's just like any . . . kind of surgery. When you got the scars, you got the scars and they're there for life. But it certainly gives me a whole new set of military experiences to look on with a great deal of pride that will supercede some of the military experiences I look on from the Vietnam days."

Schwarzkopf was indeed correct. The scars will remain not only for the life of those who fought there but in the history books as well. However, the stigma spoken of above can be purged. How? By insuring that both our political and military leaders remember the hard lessons.

It is evident that it is tragically irresponsible to send forces into battle without allowing them to exercise their full range of strategic and tactical options. They should not be committed at all unless a clear-cut victory is one of the goals.

One of the mistakes America's political leaders made in the sixties was to try and cast our military officers as efficient managers rather than dynamic combat leaders. That was the McNamara Syndrome, the theory that what worked for big business would work for the military. One had assets and one managed those assets to produce a product or service for the purpose of making a profit. In the military, so the theory went, your assets were your men and hardware; you managed them with a priority on cost-effectiveness; and your profit was the accomplishment of your mission with maximum bang for the buck.

There were several fatal flaws with the McNamara Syndrome theory. One, management cannot stand by itself in a military environment. Commanders must be leaders of men first, with management capabilities among their skills.

Two, war cannot be fought with cost-effectiveness as the driving force. There will always be waste. There is no way you can exactly forecast a campaign's or battle's progress from the standpoint of the expenditure of your equipment and supplies—and men. To insure success, you have to plan for the worst-case scenario. If the outcome is more favorable, you will have had no need for all of your resources. You have "stuff" left over, bought but not used. Maybe a business can't operate that way, but a military force must. If your study reveals that an infantryman uses an average of three bullets to kill one enemy, you don't restrict him to just three bullets. Maybe next time the enemy fights harder and with more effectiveness, and your infantryman dies because he needed four bullets.

Three, the "manager" can't sit several thousand miles away (in Washington, D.C., for example) and be as well-versed in the details of a battle as the commanders in the field, even with today's instant communications. And while Western constitutions provide for civilian control of the military (and rightfully so), they do not address civilian strategic or tactical conduct of a military campaign.

The Vietnam veterans who led U. S. forces during the Persian Gulf conflict had all learned the lessons of Vietnam (as had the civilian leaders in Washington), and will pass those lessons down to the next generation of military commanders. And the success of Schwarzkopf's hit 'em fast, hit 'em hard, and don't pull your punches campaign is proof of the pudding.

There are two other important factors to consider with respect to our "new" military.

Technological advances have dramatically altered the effectiveness of weapons, when employed by properly trained personnel. Armies are more weapon-effective now. Delivery systems have improved to the point where we can almost say "one target, one weapon." But not quite. Still, there has been a quantum leap.

The United States has fielded an all-volunteer force in battle—for the first time. The people are better educated, more highly trained, better paid, and more sincerely motivated than ever before. A military career is now a most honorable thing and as such it attracts a higher caliber of enlistee.

Also changed is the concept of reserve forces and the National Guard. No longer ill-equipped with make-do, cast-off equipment and weapons systems, the new civilian soldiers, sailors, airmen, and marines have modern equipment, and drill alongside active forces, with equal responsibilities and skills. And they quite frequently best the regular forces in exercise competitions. Such forces played a vital part in the Persian Gulf war, as most of the USAF airlift capability is manned by reservists; and several other key tasks (water desalination, minesweeping, and field medical units, to name just a few) are almost the exclusive purview of reserve forces.

General Schwarzkopf was the beneficiary of a "new" American military and he gave his forces the one thing they needed to round out their maximum capability: exemplary leadership that included political, diplomatic, *and* management skills, as well as combat savvy based on personal experience and a lifetime of practice in the military arts and sciences.

He also added one other quality: a strength of will and determination that made him certainly better than the average bear!

CHAPTER 4

IN EARLY NOVEMBER, Dick Cheney visited with Schwarzkopf on his way to confer with King Fahd. It gave the two another chance to discuss the military and political developments and Cheney used the opportunity to visit with the troops. Once more, a high official carried news of the support of the American people directly to the men and women involved, and they reacted with renewed confidence and determination. Cheney was pleased to see that everything that could be done for morale was being done. Scattered about were a number of so-called "roach coaches" that offered the men and women around-the-clock sodas and chips and sandwiches and in some cases even a non-alcoholic near-beer. Mobile and amply stocked, they were somewhat misnamed and well used, a combined snack bar and place for brief social contact.

Schwarzkopf and Cheney also discussed the command and control aspects of DESERT STORM, and one of SecDef's tasks on his mission to see the Saudi ruler was to present for Fahd's approval the chain of command for Saudi forces.

This was a delicate issue but one that had to be ironed out prior to the start of any hostilities. Saudi Arabia was the host country and the one most at risk. But it was recognized that the United States, who provided the bulk of military power, would conduct the operation, and coalition forces would be directed by Schwarzkopf but under their own commanders, in elements that conformed to their own political groupings. It is to the credit not only of Schwarzkopf but also of Powell and Cheney that when the battle did start two months later, there was a clear-cut assignment of tactical areas of responsibility and military goals. In recognition of the host country and its position in the Arab world, the Saudis were positioned nearest the front lines as the buildup continued.

Almost immediately after the U. N. sanctioned (on November 25) the use of military force if Iraq did not withdraw from Kuwait, President Bush made a surprise announcement (on November 30). He would be willing to meet with the Iraqi foreign minister, Tariq Aziz, and dispatch Secretary of State James Baker to Baghdad for talks in an "extra mile" attempt to prevent military confrontation. Within a few days—Saddam accepted the proposal on December 1—Baghdad agreed to the meetings but called Bush an enemy of God and vowed to bring up the question of a Palestinian homeland at the talks. Bush's response was that there would be no talks on the Palestinian question; the only subject on the floor was the unconditional withdrawal of Iraq from Kuwait. Period. End of sentence.

Along with daily reports that told of the seesaw progress of the impending talks, the American public was treated to a wealth of articles and pictures of Arab and Western soldiers training side by side and conducting exercises together. It was gratifying, and stimulated thoughts that when the immediate crisis was over, there could possibly be some meaningful dialogue and progress toward a settlement of the political problems that had plagued the Middle East for so long. Arabs and Westerners could work together on the battlefield; why not within the halls of the United Nations?

President Bush continued to speak of a new world order, one in which nations could finally resolve their age-old conflicts and unite in a great community that could address the world problems of poverty and hunger and environmental destruction. What a spin-off that would be from the events of DESERT STORM!

Cheney's visit also gave him an additional opportunity to evaluate with Schwarzkopf the DESERT SHIELD force buildup vis-à-vis the recent increases in Iraqi military strength. It was apparent that while Saddam continued to throw out hints of his willingness to negotiate a peaceful settlement, his real intent was to remain in Kuwait. Only concessions on the part of the coalition had any chance of reversing his stand, and President Bush was adamant that only complete, unconditional withdrawal was acceptable. It was becoming clearer that the exercise of the military option to force Iraq out of Kuwait was growing more probable with every setting of the desert sun.

More American and coalition forces were deployed. The battleship *Missouri* and its battle group departed U. S. waters for the Persian Gulf.

On November 15, Operation Imminent Thunder was ordered and U. S. Naval forces prepared to conduct a highly visible full dress rehearsal of an amphibious and airborne assault across Saudi beaches just south of Kuwait City. Certainly, such a tactic was a viable option wherein the marines would storm ashore with the same battle-wise tactics they had developed in bloody conflict down through the years. Despite a delay caused by inclement weather, the operation received extensive media coverage and there was considerable speculation by the press and television analysts (some of whom were former military men) that at the outbreak of hostilities the marines would land along the southern shores of Kuwait and strike immediately northward for control of the Kuwaiti capital. Elaborate charts were displayed and potential tactics discussed, as if the analysts themselves would do exactly that if they were conducting the operation. And to lend credence to those opinions, several other naval assault exercises were conducted during the following weeks.

In his Riyadh headquarters, General Schwarzkopf kept track of the coverage with supreme satisfaction. For the moment, things were going exactly according to his plan. In response to the operations in Saudi Arabia, Iraq shifted another 250,000 of its troops south into Kuwait and mobilized another 150,000 reserves, largely poorly trained and poorly equipped recruits. The Iraqis moved more troops to the beaches of Kuwait City and continued building an extensive land barrier of troops, razor wire, berms, and ditches, not only along the beach area, but along a portion of the southern Iraqi-Saudi border.

Schwarzkopf made his next move. He deployed additional forces to positions where they were directly aligned along the northern Saudi-Kuwaiti border, opposing the bulk of Iraqi forces. It was the beginning of a classic deception.

The canny but wary Bear knew that he still had some serious obstacles to overcome. According to doctrine, an attacking force should have a numerical advantage of 3:1 or more (5:1 for example, if you're going up against dug-in forces). He knew he wouldn't have that. At his early-August meeting with Cheney, Powell, and Bush in Washington at the outset of deployment, he had suggested such a ratio and later had considered his recommendation perhaps a bit unrealistic considering the active duty strength of U. S. troops and their other commitments. He would be doing well to have *any* significant numerical advantage. Considering the advanced state of training of his forces and the high-technology weapons he would have at his disposal, he knew he could live with less than 3:1, probably much less. But included in his troop count were the logistic support forces, who were certainly essential. But that meant his troop count, when it came to those who would actually face the Iraqi soldiers in the field, was actually less than the numbers generally announced.

The same situation would probably apply to his

armor. The Iraqis could very well have a numerical advantage. But he should have the technical advantage; the Abrams was most probably the best main battle tank in the world, and it had foul weather and night capability unmatched by even the touted Soviet T-72.

After an overall evaluation, he concluded that he would have as much as a 3:2 *disadvantage* in front-line troops and a slight disadvantage in tank numbers. If his assumptions were correct, he had no choice. He would have to make up the difference by superior strategy and tactics.

It might be helpful to define better those two terms, strategy and tactics, as they pertain to the battle commander. Basically, strategy is his overall battle plan, and tactics are his methods of accomplishing his plan. DESERT STORM was Schwarzkopf's strategic plan. The battle movements of his forces, most carried out in accordance with developed service doctrine, would constitute the tactics of the operation.

He would gain some advantage through the superiority of his men and weapons; he could gain additional advantage by deception and movement. And there was not a commander around who could capitalize on those two assets better than Schwarzkopf. Hit 'em hard. Hit 'em fast. And keep 'em guessin'. Knock 'em down and stomp 'em. Then it could be Miller time. And to help him set up the Iraqis for just such a blow, he had very powerful air and naval resources.

In fact, his staff had prepared and was constantly updating a comprehensive target list for air strikes, sea-launched missiles, and naval gunfire that, properly executed, would devastate Iraq's military command, control, and support capability, and set up her forces in southern Iraq and Kuwait for the kill. To that end, Schwarzkopf had an awesome array of weapons.

The first order of business was to gain control of the airspace over the theater of operations. Iraq had a formidable air force, but Schwarzkopf had no doubt that Lt. Gen. Horner's assets were equal to the task.

The USAF F-15 Eagles and F-16 Falcons, in concert with USN and USMC F/A-18 Hornets and coalition Hornets and Jaguars, would be given the mission of sweeping the skies of Iraqi Fulcrums, Mirages, and anything else that dared to venture forth. Navy F-14 Tomcats would also participate, as well as maintain their role as fleet defense fighters to cover any over-water forays by Iraqi aircraft.

With air superiority, coalition jets could go after defense radars and start their attrition of command and control facilities, communications centers, and logistic supply lines.

For that purpose USCENTCOM had at its disposal the old but very effective F-4 "Wild Weasel" Phantoms who could detect antiaircraft fire-control radar and reply with beam-riding ASMs to destroy the detected radars. They would also be used for sophisticated radar-jamming missions which would effectively blind any remaining sites. An array of aircraft would be available for direct attack of facilities. In addition to the fighters mentioned above, there were the F-15E Strike Eagle, a two-seated fighter-attack bomber, and the extremely low-flying and ultrafast Tornado manned by crack British and French pilots. The long-range F-111 Aardvark, so named because of its slim pointed snout, could sniff out and destroy a variety of targets while its team member, the EF-111 Raven, would handle electronic jamming and other suppression duties. The Navy's all-weather, 24-hour A-6 Intruder, paired with its sister aircraft, the EA-6B Prowler, had a one-two punch that would simply be irresistible by the Iraqis. As if all that were not enough, the state of the art F-117A Stealth tactical fighter was also capable of bombing and missile attack missions. Marine Corps VTOL (Vertical Takeoff and Landing) AV-8B jump-jet Harriers were also in the inventory but their major role was air support for marine ground forces; still, they would be employed in attack missions.

Also available, but destined primarily for the ground attack phase, were the AH-64 Apache, the army's battlefield antitank and attack helicopter, and the army's and marines' AH-1 Cobra and Sea Cobra.

This formidable air force was capable of being controlled by both ground and air command-and-control centers with around-the-clock employment of the large E-3 AWACS Sentry (a military version of the civilian Boeing 707 airliner loaded with electronic detection, control, and jamming gear), and the smaller, carrier-based E-2C Hawkeye, the Navy's similarly tasked aircraft.

Offshore, waiting patiently, were the battleships *Wisconsin* and *Missouri* with their 16-inch guns and Tomahawk cruise missiles.

For the final phase of the softening-up campaign, giant B-52s were ready as required in England and stationed off Africa on the island of Diego Garcia.

The thunder was almost ready and the lightning was building in the wings.

Thanksgiving was approaching, and that is always a hard time for service personnel away from their loved ones—as it is for those they left behind. For several days before and after, there is an emotional letdown as memories of past celebrations at home remind everyone of the sacrifices they are making. This is partially offset, however, by the special efforts made to provide traditional meals and expedite delivery of the many packages from home. Makeshift decorations always seem to appear, and along with the sorrow of separation are thoughts of the real meaning of the holiday.

Schwarzkopf had much to be thankful for. He wasn't in a shooting war, yet, and thus could celebrate a peaceful meal with his troops. The buildup of his forces, the largest military movement since that which preceded the Normandy invasion of World War II, was proceeding with a minimum of setbacks. He was at the peak of his career. His family was well and thinking of him, and there were a million memories of them to help him through the occasion. The traditional sit-down family dinner was one such memory. Having time to just sit after the meal and scratch behind the ears of his dog, Bear, was another. Like so many of us, Schwarzkopf probably remembered the pleasant discomfort of a too-full stomach.

And this year, the president and his lady were coming for dinner.

On November 20, President and Mrs. Bush arrived in Saudi Arabia and enjoyed the holiday meal with American forces. Bush understood the dangers facing the men and women as well as any president ever had. Bush had been a young naval aviator during the latter years of World War II and his torpedo bomber had been shot down in the Pacific. He had survived and been rescued by an American submarine that completed its wartime patrol to Tokyo Bay before returning Bush to safety. Many of the men and women who had the opportunity to speak with him during his Thanksgiving visit knew of his experience, although it may not have been general knowledge among the troops, and there was a special feeling of appreciation between the two generations, each recognizing the other's sense of duty.

Bush brought with him the good wishes of the American people and the confidence of his office and Congress that the forces of DESERT SHIELD were up to any task that faced them.

The end of November came and passed. The desert nights were cool and sometimes cold. It was now the middle of the fall season and back in the States it was time for football, when every weekend millions of Americans were either in the stands or sitting before their TVs. As they watched, they noticed something new on the helmets of the professional teams, a small American flag. The NFL was showing its support of the U.S. commitment in the Persian Gulf area. When the pre-game national anthem was played, people stood a little straighter, and many more removed their caps and placed them over their hearts. Most joined in the lyrics. A lot of eyes were moist. Life was going on as usual but they were remembering.

Bob Hope, the perennial Christmas entertainer for every war since the American Revolution (or so it seemed) was making plans and putting together his routine and assembling his company. This would be a different visit. There would be pretty girls, but there would be no skimpy costumes out of respect for Islamic customs. The jokes would have to be selected with the same consideration. But the show would go on and it would bring a few minutes of escape, as it always had since the dark days of World War II to military men and women. The Les Brown band would play and after an introduction by the base commander, Hope would saunter out from backstage, golf club in hand, with a unit ball cap perched cockily on his head. His advance men always provided him with some in-house humor and as always he was an instant hit. The old trick of inviting a member of the military audience up on the stage to be sung to by a pretty girl seemed fresh to the young troops of DESERT SHIELD, and they parroted their fathers and grandfathers in hooting and hollering and whistling as the embarrassed selectee rolled his eyes, nervous at being held so close to a woman in full view of several thousand of his buddies. Afterwards, Hope would join his female singer and they would hoof through a short soft-shoe routine and he would launch into another brief monologue. No matter that the same jokes had been told fifty years ago and only the names had been changed. Laughing at Hope was almost a war ritual, like cleaning your rifle before battle or sticking a cigarette pack in your helmet strap.

Finally, someone would sing "White Christmas," perhaps Dolores Hope, Bob's wife. The sight of thousands of young warriors sitting quietly in the sand, respectfully listening to an elderly lady softly singing (slightly out of pitch at times) the classic Christmas melody was a bit of Americana. Then the band would lead in with "Thanks For the Memories," Hope and company would wave good-bye, and the troops would go back to the sand and the lizards and the big black beetles and the scorpions and the two kinds of very poisonous snake that called Saudi Arabia their home.

Hope has a real love for the troops and they respond in kind. Later, those back home would relax around their television sets and watch the replay, their eyes searching the frequent shots of the audience for that one special face.

Over the last half-century, Bob Hope has been in more theaters of operations than any grunt, stood on more tarmacs than any zoomie, trod on more decks than any swabbie, and sloshed his way though more jungles than any jarhead. It is a pleasure to recognize him here, in this story of a general who also places his men's morale so high on his priority list.

The stars over the Middle East at Christmastime were at their brightest and the troops of DESERT SHIELD/STORM slept restlessly through those nights, wondering what was to come and where they would be next Christmas.

CHAPTER 5

AS CHRISTMAS PASSED, Mrs. Schwarzkopf decided to leave up the decorated tree and, under it, the "big" presents for her husband to enjoy when he came home. Like all service wives, she was very concerned about his safety although she was aware that most of the time he would be at his headquarters. Still, there were unknowns and she worried about some of the same ones that her soldier did. What was Saddam going to do? On December 22, he had threatened to use chemical weapons. And if he did, would his aircraft or missiles or artillery reach into Riyadh? They could. She knew that. All of the same fears she'd had when her husband was on his second Vietnam tour returned—with a new one: the terrible prospect of chemical and biological warfare. It didn't take much imagination to envision a rain of such weapons falling on USCENTCOM forces, though it was hard to imagine just what kind of madman would do such a thing. Obviously, Saddam had done such a terrible thing before.

Schwarzkopf had the advantage of a military man's knowledge of such weapons. They *were* terrible, but their effect was mostly psychological if you were prepared for their use—but at best, the combat efficiency of the protected soldier would be affected.

As previously stated, alert forces could counter a chemical or biological attack with considerable effectiveness *if* they acted in time. But that was not really a practical assessment of the threat. There could very well be surprises, and the agent could be dispersed before the men and women had time to don their protective clothing and masks. Arms and equipment could become contaminated and decontamination made difficult. Behind-the-lines terrorists or battlefield infiltrators might clandestinely deliver the weapons, which could be catastrophic. Schwarzkopf felt that his forces were capable of defeating the physical threat and prayed that they would have the opportunity to know that it was coming. All they needed was a few minutes' warning. But there was another disturbing factor. American forces were equipped with an old-style mask. It was a capable piece of equipment, but the filters were not quite as long-lasting as those on newer models used by some of the coalition troops. There was another drawback: To change the filters required removing the mask, and it took several minutes to complete. It would have been nice to have had masks with better peripheral vision and better voice transmission capability.

In addition to the chemical agent, mustard gas, that produced horrible burns both externally and within the lungs if breathed, there were several potentially lethal biological agents. These could be delivered by a variety of means: the warheads of conventional ground and air weapons, for example, although it appeared these would deliver smaller doses than Saddam would have liked. Nevertheless, there's no such thing as a "little bit of biological agent," any more than there is "a little bit of pregnancy."

There were several commonly known agents:

Anthrax, a bacterial disease of cattle and sheep but also contagious to humans, had been around for a long time and, like the other biological agents, it infected its victim primarily by inhalation. The result is fever, acute respiratory distress, and shock.

Botulism, a very familiar food-poisoning bacteria caused by anaerobic soil bacilli, causes dizziness, paralysis, and respiratory failure.

Plague, a terrible microbe that also causes fever and acute respiratory distress.

Tularemia, a deadly bacteria that, again, manifests itself by fever, chills, and acute respiratory distress. A tiny amount of this—less than a gram—could cause thousands of casualties.

All of the above could most effectively be delivered as a fine mist (perhaps as from a group of bomblets), but could also be introduced by deliberate contamination of the environment. There are antidotes and each American soldier carried syringes in the pockets of his protective suit for self-injection.

Even untreated, there is a 15 to 40 percent survival

rate, depending on the agent used and the health of the infected victim. Those, of course, are not comfortable figures and certainly not acceptable. Even battlefield survivors would be rendered incapable of fighting. A few of these agents contaminate an area for a long time; thus Iraqi troops might have some hesitation in using them on their own soil. Also, it is possible that new strains of any of the above could be developed that would resist inoculation. Such a threat could not be downplayed by a prudent commander. And the threat applied equally to ships at sea and behind-the-lines airfields and command posts.

A few doom-and-gloomers back in the States and Europe, without any real knowledge of the agents or the military's ability to counter them, predicted widespread disaster if they were used, and such irresponsible proclamations caused considerable distress to loved ones back at home. Concern was certainly appropriate, but the public needed enough information to keep chemical and biological warfare in the proper perspective.

Unfortunately, nuclear attack is not in the same league, and is a very real threat to be seriously feared. Saddam did not have that capability, but he was progressing toward production of nuclear weapons and there seemed to be no doubt among those who knew his record of murder, assassination, and attempted genocide, to think he would not use such weapons if and whenever he got them.

Early January was a critical time. The January 15 deadline set by President Bush for the withdrawal of Iraq from Kuwait crept closer with each breath. Iraq and the United States were publicly reinforcing their intentions.

"Should the Americans become embroiled," Saddam had threatened, "we will make them swim in their own blood, God willing."

"There are times in life when we confront values worth fighting for. This is one such time," Bush stated. He gave voice to the feelings of the allied coalition as well as the vast majority of his countrymen. Polls showed that war, as much as it was dreaded, was seen by the American and coalition publics as an acceptable solution to an otherwise unsolvable problem. Not every citizen agreed on that, of course, and groups opposed to such a path were active and vocal. Thousands had taken to the streets in San Francisco, Washington, Tokyo, London, Bonn, and Paris, but crowds were nowhere near the size of those in the sixties. It seemed that every city of any size had a few who felt that no goal, no responsibility, was worth fighting for. It was unfortunate that they were unable to communicate that message to Saddam. The very small opposition was nothing like the scale of discontent people voiced over U.S. involvement in Vietnam. Nor should it have been.

Still, the reports of peace marches and protests reached the troops in Saudi Arabia, and their reaction was as selfless as their service. "That's their right, but I don't agree with them," one soldier was quoted as saying. "That's one of the reasons we're over here," said another, lifting his shoulders in resignation. Others were less sympathetic and a few must have wondered what it takes for some people to support their country's difficult international decision. Here we had a naked, brutal takeover of an independent nation by a ruthless dictator who demonstrated little regard for human life outside of his own. He had used poison gas on his own people during the Kurdish uprising, in July of 1988. The whole world had risen up in alarm and anger. If ever there was unity of purpose and a just cause for action, it must have been now, with the world community pledging to remove the Iraqi invaders from Kuwait and return the tiny nation to its legitimate government. America should be proud in taking a leading role.

There was still a procedural disagreement within the U.S. Congress. All along there had been off-the-record opinions and public statements on the legality of Bush's actions in activating DESERT SHIELD. Some members of Congress felt that the authority to go into battle was a matter of declaring war, and that was the Congress' prerogative. Saddam Hussein, ignorant in the ways of democracy, took such debate as indecisiveness. He believed there was a national lack of will. He even publicly thanked the war protesters, completely misreading their importance and influence. He boasted that, should the American-led forces come at him, "The Americans will come here and perform acrobatics like Rambo movies. But they will find here real people to fight them. We are a people who have eight years of experience in war and combat."

Finally, in the last days before the deadline of January 15, the U. S. Congress took up their formal debate as Speaker of the House Thomas Foley reminded them that "this is a matter of enormous moment." Within hours, the Congress voted to approve and stand behind Bush's decisions. Indicative of the bipartisan mood of

Congress, once the resolution had passed, was the comment by Wisconsin Democrat Les Aspin, the chairman of the powerful House Armed Services Committee: "If all else fails, war is a reasonable option."

The President had the mandate of the American people, its senators and representatives in Washington, and the military forces in the field. The die was cast, and only Saddam could stop its roll before the United States made its point.

Schwarzkopf and Company watched the events back home, always hoping that there would be some kind of acquiescence by Saddam. But in a reversal of his previous agreement to confer with Baker in Baghdad and send his foreign minister, Tariq Aziz, to meet with Bush in the "extra mile" talks, Saddam suddenly declared that he could not accept any of the dates proposed by the United States for his conference with Baker. Baker and Aziz did meet in Geneva, however, and Baker attempted to convey a letter from Bush through Aziz to Saddam. It read in part:

> There can be no reward for aggression. Nor will there be any negotiation. Principle cannot be compromised. However, by its full compliance, Iraq will gain the opportunity to rejoin the international community.
>
> More immediately, Iraq and the Iraq military establishment will escape destruction. But unless you withdraw from Kuwait completely and without condition, you will lose more than Kuwait. You may be tempted to find solace in the diversity of opinion that is American democracy. You should

> resist any such temptation. Diversity ought not to be confused with division. Nor should you underestimate, as others have before you, America's will.

Aziz refused to accept the letter for delivery to Hussein, saying he was not authorized to accept it.

With respect to the proposed Baker-Saddam talks, Saddam insisted he was just not available until after January 12. Bush felt the twelfth was too late a date for the withdrawal to be commenced and be completed by the deadline. Saddam wanted three weeks to withdraw. Bush declared one week was sufficient, and that was a provision of the withdrawal demand.

Two things were apparent. Saddam wanted the meeting to be on a date of his selection to maintain some control and stature within the Arab community, and it would amount to a backdown of the United States regardless of the outcome. Bush stuck with his one-week-to-withdraw provision to insure that while Iraqi troops could make it out in that time, they would have to leave much of their warmaking hardware behind.

The "line in the sand" that had been drawn back in August by the same stick that had spelled out unconditional withdrawal as the only acceptable solution to the crisis was getting set in concrete.

General Schwarzkopf had his ground forces in position and he was prepared to order the air phase of DESERT STORM anytime after the January 15 deadline. His troops were at their maximum strength, with a slight numerical deficiency over the dug-in Iraqis, but Schwarzkopf counted on deception and surprise to offset that. He knew that the French Legionnaires over on his western flank were as anxious as their American comrades to get on with the show, and the commander of British forces, Lt. General Sir Peter de la Billiere, had assured him that the Brits were also champing at the bit and eager to get things on. Sir Peter, the most decorated soldier in the British Army, had spent 15 years in the Middle East area, and had fluent command of the Arabic language. In effect, he was on "home turf." He wore the prestigious Military Cross for bravery in battle. Not only was he Schwarzkopf's fellow battle commander, the two had become personal friends. Such rapport was invaluable when in combat.

The coalition had overwhelming superiority in air power. Coalition armor was ready and technologically superior, although Iraq had 4,000 tanks to the coalition's 3,500.

Schwarzkopf knew only too well the edgy, nervous feeling in the stomachs of his people, but that was normal and would pass. They were ready and they knew it.

There was a soldier's practicality in the way they waited. Many sent home souvenirs they did not want to carry into battle. Some were writing letters even if they had just sent mail home a few days before. Their weapons were clean, condoms stretched over the barrels of the rifles to keep out the gritty sand. Tankers were policing up their tanks, getting rid of any loose or flammable materials not absolutely necessary.

Marines were sharpening their fighting knives and checking their personal equipment as they had done since the halls of Montezuma and the shores of Tripoli.

Pilots were oiling and checking their handguns and making sure that nothing but necessary items were in the pockets of their flightsuits. No need to let any unofficial ID or personal letters fall into Iraqi hands if they had to punch out and ride the nylon escalator down.

The Navy was on station in the gulf with six carrier battle groups, the battleships *Wisconsin* and *Missouri*, an amphibious task force with embarked marines, a hospital ship, and sufficient small combatants and support ships to sustain operations as required. In addition, two Marine Expeditionary Forces were in forward ground positions alongside the army, and coalition forces were strung along the entire Kuwaiti-Saudi border.

January 15 came and passed.

The world sat on the edge of its seat.

DESERT STORM

Thursday, January 17, 1991, 2:40 AM, Iraqi time
Wednesday, January 16, 1991, 6:40 PM. EST
(Except where noted, all dates/times following are Iraqi time)

CHAPTER 6

THE METAMORPHOSIS WAS ALMOST instantaneous.

Overnight, the great, growing caterpillar of allied air forces in the desert of Saudi Arabia, joined by ground-hugging Tomahawk cruise missiles launched from warships in the Persian Gulf, burst forth as a lean and mean fighting machine, a deadly multinational butterfly streaking determinedly through the moonless night sky over Baghdad to deliver the first devastating blows against the military capability of Iraq. Their concerns the inverse of moths attracted to a flame, the attacking aircraft evaded a sky full of triple-A (AAA, antiaircraft artillery) to deliver the firm message of the major Western powers, the Arab nations, and practically all of the members of the United Nations: Withdraw from Kuwait.

DESERT STORM was under way.

Caught by surprise despite a five-month buildup of coalition forces and repeated warnings that military force would be used anytime after the midnight, January 15 (EST) diplomatic deadline, Iraqi forces reeled from the ferocity of the first attacks and never recovered. Battle-hardened, but sadly misled by the desultory pace of eight years of crude conflict with Iran, Iraq simply had no understanding of the overwhelming military technology of the forces arrayed against it. *Only a determined allied effort* to keep both coalition and Iraqi casualties to a minimum allowed Iraq to escape overnight annihilation. A combined all-out air, sea, and land assault could have erased Iraq as a nation and as a culture. But the aims of the allied nations, as steadfast as they were, and as directed by the commander of DESERT STORM, General H. Norman Schwarzkopf, did not include the indiscriminate and unnecessary taking of life. The conquest of Iraq was not even a goal; the restoration of the sovereign state of Kuwait was the only objective. And the first strike in the darkness of the early morning of January 17, 1991, was the beginning of the military operations aimed at achieving that objective.

The devastating air phase was orchestrated by Schwarzkopf's air component commander, Lt. Gen. Charles Horner, USAF, who wore a second hat as the coalition's air commander. Horner and his number-one targeter, Brig. Gen. Buster C. Blossom, USAF, had set up shop in a basement office of the Saudi air force headquarters. Super-secret, the targeting space was accessible to no one without a strict need-to-know, and was soon dubbed "the Black Hole." Those inside spent the entire war insuring that coalition aircraft were assigned appropriate targets and missions. It is a tribute to their professionalism that the air war, consisting of tri-service and coalition forces, was run by one commander in a precedent-setting example of air command. So classified and so thorough was Horner's "Black Hole" team, it was not even reported on until after the cease-fire.

From the first burst of a plummeting bomb and the somewhat tardy reply of Baghdad's anti-air defenses, the news media were at the obvious scene of the action and reporting live.

Although foreign correspondents and reporters of all three major American networks were present, the scoop of the new decade went to CNN, the Cable News Network. A worldwide all-news station, CNN had been structured for just such an opportunity: around-the-clock war coverage.

Back in the States it was the dinner hour (early evening, January 16) and a large segment of the viewing audience was mesmerized by the voices of three CNN correspondents—Bernard Shaw, Peter Arnett, and John Holliman—as the men crouched in their room at the Al-Rashid Hotel and over an uncut telephone line gave a vivid eyewitness account of the air attack on military installations within and on the outskirts of the city. Their report was relayed over CNN's television audio network while viewers watched a still picture of the faces of the correspondents. No one who listened will soon forget the trembling apprehension in the three men's voices as they described the frightening details of the attack. Justifiably concerned for their own safety, the trio nevertheless held their posts, sometimes lying low on the floor while holding their microphones out the window to record the booming and shattering noises of death. Their verbal reports reflected a devastating drama in progress and when their voices occasionally cracked, we who were with them via the magic of satellite and telephone communications also sucked in our breaths and felt our hearts stand still as uncertainty seemed to precede certain catastrophe. Yet their hotel was not hit—and that was no accident. The first attack of the war was a precise surgical attack on selected military targets, and for the first time the predominant weapons were the "smart" kind with an uncanny ability to hit either directly or within a few feet of their intended target.

One of our first emotions had to be awe at the superb performance of high-technology weapons. Almost immediately, the world knew this would be a new level of an age-old rite. Since the first time an *Australopithicus* swung his tree-branch club at the head of a fellow prehistoric man, individual and collective combat had progressed (if such a trait can be called progress) through the rock, the spear, the bow and arrow, the flintlock, the cannon, the rifle, the automatic weapon, the iron bomb, and the nuclear follow-on to weapons that seemed almost to have a mind of their own. Using preprogrammed computer guidance that compared stored data with radar returns of actual terrain, the Tomahawks, for example, could precisely navigate, and even go around corners to strike with deadly accuracy their preset targets. No longer did attackers target a bunker; they targeted the front door. To destroy a building they guided several hundred pounds of high explosive down a ventilation shaft. To counter an incoming missile, they fired another missile that homed in under the guidance of a tiny cathode ray image of the attacker and met it head-on high in the sky. The three CNN men crouching low in their room at the Al-Rashid were sending out the first combat reports of the world's premiere high-tech war. True, earlier versions of some of the weapons had been used in the Vietnam conflict, but the Persian Gulf would be the first conflict with massive employment of highly improved, computer-controlled weapons. They were the mainstay of land, sea, and air attacks.

In all probability, Saddam had no real feel for the capability of the forces aligned against him. But one assumes that lack of understanding quickly disappeared on the morning of January 17. DESERT SHIELD became DESERT STORM and from the skies over Baghdad came the first explosive signs of the beginning of the end for Saddam Hussein.

At that same moment, the television media as represented by CNN, had one of its finest hours. Regrettably, that same network later became the object of considerable criticism as it embarked on a path of reporting that many saw as serving the interests of the enemy.

When video was available from Baghdad, viewers around the globe saw the dark early-morning skies over the capital city laced with thousands of rounds of yellow and red antiaircraft fire that rose as if in slow motion and then curved down toward earth as each shell reached the apogee of its deadly search for targets. The world got its first eyewitness accounts of the accuracy of allied weapons as the CNN trio described the military targets hit and the obvious civilian ones that were being spared. As the attack wore on there was a definite change in the tone of the reporters' voices as they expressed relief that their hotel appeared not to be on the target list. (Earlier, it had been, as there was a suspicion that an Iraqi command and control facility was in the basement; but with the hotel full of international correspondents, it was removed.)

On the horizon of the city, great orange balls of light burst and quickly faded, transient testimonies to the powerful explosions of allied weapons. Wailing sirens, the staccato voices of antiaircraft rapid-fire weapons, the *wham-wham-wham* punctuation of larger caliber but slower firing guns, and the earth-shaking *whump* of exploding bombs all combined to provide a background score to the start of a real-life miniseries. No special effects could ever equal the awesome display of death and destruction that began only a few hundred yards from the hotel window and extended over the entire city. True, the allies were concentrating on military targets and military personnel manning those installations were bearing the brunt of the casualties; but, as in all battles, nonparticipants would also suffer and die. It is an unavoidable part of the horror of war.

Overhead, sometimes heard but largely unseen, coalition pilots were racing across the skies, their eyes glued to their heads-up displays (HUDs) and the cockpit portrayals of radar- and thermal-sensitive sighting devices. Their hands gripped their flight controls with the firm but slightly moist grasp of professionals on an adrenaline high. Those on low attack profiles used their speed and abrupt maneuverability (jinking) to foil the detection and lock-on attempts of the angry gun crews below. By the time they could be seen or heard and the guns positioned to fire, they were gone.

Several sinister, triangular, ink-black fighter-bombers slipped through the dark sky completely undetected, and one placed its weapon directly on a Baghdad telecommunications center. The F-117A Stealths had flown their first combat missions.

As the attack went on, one startled British television reporter standing on a sixth-floor balcony of the Al-Rashid Hotel watched a Tomahawk cruise by at his level and crash at full speed into the Iraqi Defense Ministry building.

Baghdad was not the only focus for the initial attack. From northern Saudi airbases and a major airfield at Dhahran, aircraft from the U.S., Britain, France, Canada, Italy, Saudi Arabia, and Free Kuwait fanned out across Iraq and occupied Kuwaiti air space. Aircraft from the Navy carriers in the Persian Gulf and Red Sea had joined the attack from its outset.

High over the sands of the Arabian desert, air control AWACS and Navy Hawkeyes monitored the flow of aircraft to and from the targets and provided control where required.

Primary targets were command and control facilities, airfields, air defense installations, SCUD launch sites, and chemical/biological production plants. Coalition pilots were also prepared to take on the Iraqi air threat, but very few of Saddam's fighters rose to challenge the allies, even after daylight came. Perhaps a number had been caught in the hardened airfield bunkers. Certainly, most of the runways had been cratered.

Schwarzkopf and his staff had assembled in his underground war room with a chaplain shortly before the battle commenced and had prayed together for the safety and well-being of the men and women of DESERT STORM. Now, with silent prayer, they studied the sketchy reports that were coming in. Reliable damage assessment would have to wait until pilots returned and were debriefed and satellite photographs could be examined. The CNN crew at the Al-Rashid was continuing to broadcast, however, and some information could be gleaned from their reports. The general was pleased to see that the attack on Baghdad was truly the surgical strike that had been planned. It would be a fairly safe assumption that the same degree of precision was being demonstrated in the other attack areas, though there are always doubts until some hard intelligence is received. There were no early reports of plane losses. That was phenomenal. The antiaircraft fire had obviously been heavy at times, but there had been no air opposition. Was Saddam holding back his resources?

The same success continued during the daylight hours, and coalition aircraft flew 2,000 sorties during the first 24 hours.

As pilots returned and made their reports, it was evident that they had achieved outstanding success, even though some had been unable to positively identify their targets and had been forced to return with their weapons, one of the rules being to avoid any random bombing that could wound or kill civilians.

President Bush appeared on television from the White House Wednesday evening, January 16 (U.S. date), and announced the start of Operation DESERT STORM to the American people. Similar announcements were made worldwide by either national leaders or news anchors.

"We will not fail," proclaimed Bush.

Saddam had threatened to strike Israel as one of his first reactions if attacked and early Friday morning the bad news came. Although coalition aircraft had particularly sought out Scud launching sites and had reported great success, they had been unable to get them all, especially the mobile Scud launchers that could be easily hidden during the attack and then rolled out to send the ugly missiles on their way.

The Iraqi madman did exactly that and in the early morning hours of Friday, January 18, the air raid sirens sounded over much of Israel. Israelis hurriedly donned gas masks and entered sealed rooms. It was a very uncertain time.

Scuds dropped onto Tel Aviv and Haifa. There were reports of poison gas warheads exploding which, fortunately, proved to be false. But Saddam was making good on his terrible promise. As Israeli rescue teams combed through the wreckage of destroyed buildings, Saddam announced that "The mother of all battles is under way."

Israel had earlier given assurance to Washington that they would not attack if such a thing happened, but that night a great roar resounded over Tel Aviv as Israeli fighter and attack aircraft lifted into the sky. Certainly, the terrorist missile attack cried out for retaliation. But the Israeli casualties were light, despite a total of eight

Scuds on Tel Aviv, Haifa, and Ramallah, and it was not the massive attack everyone had feared. With much gritting of teeth, Israel held herself in check, reserving the right to strike later at a time of her own choosing.

Schwarzkopf was relieved to see such restraint. An Israeli attack on Iraq could have caused the Syrians and Egyptians to have second thoughts about remaining with coalition forces. It was widely anticipated that the two Arab countries would not fight *with* Israelis against fellow Arabs.

On the same night, an historic event took place. A Scud was fired at Saudi Arabia, but U.S. Patriot anti-missile missiles roared off their launchers and destroyed the Scud in flight, the first time one missile had destroyed another in midair. It was an instant vindication of the years of effort and millions of dollars spent developing the Patriot. The American SDI program immediately received a boost from the success of the very high-tech weapon.

Television reports showed coalition aircraft beginning to return from strikes, the jubilant pilots climbing down from their cockpits, and giving each other "high fives." The morale of the flight and ground crews soared sky-high.

Four British pilots were interviewed right after dismounting from their Tornados and a squadron leader reported on their mission with the usual English reserve. Yes, yes, there had "been a bit of flak" and the mission "was a rough go for a while." Two of the other pilots quietly agreed and allowed as how they, too, were pleased with the mission's success. You would have thought they had just returned from an "acrobatic do" over the Thames. The junior pilot, however, exhibited considerably more excitement and his voice was a good octave above the others as he told his tale. It had obviously been his first combat mission and a very rough one at that: a night attack over a strange country under heavy enemy fire. He had found his target and dropped his stores but then turned his immediate attention—as we all would—to "making a brave retreat" out of there! Bless the sturdy Brits.

With sundown on Saturday, the 19th, Israel and Saudi Arabia braced for a possible second attack. The Scud was no longer a militarily effective weapon. Its design originated with an evolution of the old German WWII V-2 by the Soviets who had intended it to carry a nuclear warhead. It didn't have to be pinpoint accurate! The Iraqis got it from the Soviets and made some modifications, the latest of which gave it (the Scud A1-Abbas) a decent range, 540 miles. A ballistic weapon, its accuracy depended on its trajectory, and that in turn was a factor of launch azimuth, firing angle, propulsion, and wind effect during its flight path.

Saddam launched three more Scuds at Israel on Sunday morning, January 20, but once again, casualties were light with no fatalities. The United States responded to Israel's request and sent in Patriot units with GI crews and they went operational in just a few hours, the first time that American troops had been deployed on Israeli soil for the defense of that country.

The air war had continued around the clock with several thousand sorties per 24-hour period. The luck of the first night had not held and a total of six U.S., two British, one Italian and one Kuwaiti aircraft had been lost, with nine American crewmen, four British, two Italians and one Kuwaiti listed as MIAs.

There had been some minor ground skirmishes between Iraqis and marines along the Kuwaiti-Saudi border and there was a report of light casualties.

The Iraqis had set up antiaircraft weapons on several oil platforms off Kuwait, and on that same Sunday the U.S. frigate *Nicholas*, supported by Navy attack helicopters, assaulted one of the platforms, taking the first enemy prisoners of war (EPWs).

Despite military restrictions on the movements of individual media correspondents, there was a saturation of coverage, particularly by television. While individual stories may not have been all the media people would have liked to have filed, polls indicated that most people were quite satisfied with the coverage.

USCENTCOM held daily operation briefings and Schwarzkopf's principal briefer, Brigadier General Richard Neal, USMC, USCENTCOM's deputy director of operations, became a familiar face to television viewers. An articulate, all-business marine with a trace of New England accent, Neal presented what factual information he could from daily briefing sheets and then opened the session to questions. To inquiries about the number of enemy casualties, he reiterated the firm decision of Schwarzkopf. There would be no body counts. Schwarzkopf had been through that routine during his Vietnam tours, and was still incensed about the inaccuracies and misleading conclusions such a means of measuring battle progress provided. During Vietnam, the national authorities thought it was the only way to indicate victory on the battlefield ("we killed more than they did"), and the count was often inflated by commanders in the field.

As Schwarzkopf would say, such use of inaccurate data was pure "bovine scatology." To begin with, there is no way that troops engaged in combat can come up with an accurate body count. There isn't the time nor is it a priority. Even if there is a fairly accurate estimate of fatalities, such information is no clear indicator of victory or defeat. If a 1,000-man force overruns a 100-man position and suffers 300 casualties but kills all the defenders, the body count indicates a 3:1 loss ratio and may be interpreted as a failed attack unless the full circumstances are known. A body count is a morbid way of reporting even war news, although some justification can be made for it, if it is accurate and it is necessary to explain a situation that the civilian community should be made aware of.

This brings us to the next dilemma that faced the media people. They did not like to trust the military to determine what events were need-to-know for the general public and yet did not have the military expertise to make that determination themselves. Many of them considered themselves military experts and a number had military service in their background, but none were currently professional officers engaged in combat and responsible for the lives of their people. And none were privy to the numerous factors, many of them changing constantly, that affected the progress of the battle.

The problem was not their intentions. The members of the fourth estate are one of the bulwarks of freedom and democracy as we know and practice it. But the operative word in the previous sentence is "one."

Enemy intelligence is always looking for the piece of the puzzle that completes the picture. And no responsible correspondent wants to provide that missing piece, even inadvertently.

We all ask dumb questions at times and correspondents are no exception. Neal listened patiently to the rare examples and usually responded with a trace of a tolerant smile, but he treated each question seriously and soon developed a rapport with the correspondents that enabled the briefings to accomplish their purpose. His serious treatment of all questions did gain him credibility with the correspondents; if he had no knowledge of a subject, he admitted it and often stated that he would look into it and get back to the questioner—although the same rules of need-to-know governed his subsequent responses.

There were other daily briefings, notably by British and Saudi spokesmen. And there were untelevised background briefings for the correspondents.

The media were never completely satisfied, which is probably one of their strengths. However, some of their conduct and statements during the DESERT SHIELD/STORM period, and the critical reactions of the public regarding that conduct and those comments, indicated that a healthy discussion of media performance, privileges, and *obligations* would be an appropriate action in the near future. And at the risk of repetition, a thorough read of Article III, Section 3, of the U. S. Constitution would be in order.

Along with the deployment of conventional forces in early August were teams of special forces: land and air commandos (Green Beret and DELTA forces) and Navy

SEALS. The missions were highly classified and their presence was not openly acknowledged, for their tasks would take them behind enemy lines and the success of their operations depended on their clandestine injection (frequently by night parachute jumps) into southern Iraq and Kuwait and their ability to avoid detection. They were truly stealth forces and they went about their business with confidence. Some Iraqi movement could be detected and reported on special radio transceivers that sent out multidata in microseconds. They used unconventional vehicles such as customized dune buggies and "dirt" bikes that would be the envy of every sand-blaster in the world if one could ever get his hands on one. Welded together with rigid roll bars and equipment lashed to every piece of body frame, the open-seated, desert hot rods with special, extremely sound-efficient mufflers raced across the sands under cover of darkness. Potential air targets were scouted and defenses determined.

Navy SEALs also surveyed the Kuwaiti beaches, took stock of the defenses, and prepared to clear the areas should an amphibious operation be required.

Special teams of Army airborne and Air Force snoopers listened in on Iraqi military communications.

As the first four days of DESERT STORM came to an end, Schwarzkopf was very proud of the coalition forces and the professional manner in which all had carried out a most demanding air assault. There would be many more days to come but they were off to an excellent start.

The beginning of the air campaign also meant that he would get out of his basement war room and personal quarters even less, and when he did venture forth he had his own personal bodyguards. But the war room was his battle station and he was surrounded by the trappings of a battle commander.

Armed guards were always right outside the spaces. Situation maps were on the walls. Representatives of each of his staff departments and component commanders had their own places to sit and perform their duties and respond to his queries or orders. There were a bevy of phones, including a field telephone with which he could reach any of his tactical commanders. Prominent was the red phone to Washington with its array of buttons. He talked with General Powell at least daily on that phone, and sometimes as often as three or four times in a day. The line to his commander in chief was always available but it would be used on only a few occasions. Inside the war room there was no telling if it were day or night, an almost irrelevant fact; war is a twenty-four-hour-a-day business. One's body clock takes a beating. Meals come and go almost unnoticed and are generally consumed without any conscious appreciation of the food, although a particularly good one is a welcome treat.

Schwarzkopf had set up a briefing and work routine for each day. But the routine quickly vanished as the demands of changing priorities set their own schedule—and each day brought with it a unique set of problems to be discussed and solved. Each day required a new consideration of orders given and orders to come.

There had as yet been no surprises beyond the light resistance the air attacks were receiving and the low numbers of casualties. Even the most optimistic estimates had been higher than the numbers so far.

The coalition pilots were itching to take on the Iraqis in the air. After all, air supremacy was the first stage of control of the airspace and one couldn't gain air supremacy if the enemy refused to come up and fight. Or could one? Schwarzkopf had figured that Saddam might hold back his response and try and sit out the first days. Perhaps that was exactly what he was doing. It might be that he thought he could wait it out, that the coalition would not be able to sustain such a pace—much like a boxer takes punishment in the early rounds to tire out his opponent. If that was what Saddam believed, he was keeping his string of miscalculations intact.

There was little probability that the air attacks would generate an Iraqi ground response. Even if they did, coalition ground forces were cocked and ready. So far, the Scuds had carried conventional warheads. Was that any kind of indicator? Was Saddam leery of using the chemical/biological weapons he had so loudly touted? Even the inaccurate Scud could be placed within the limits of a city or military assembly where exposure to the poison would be concentrated. Or had Saddam gotten the message that there would be a fearful retaliation if he used chemicals?

On the plus side there was the phenomenal success of the high-tech weapons. Initial reports indicated that of the first 54 Tomahawk cruise missiles launched from the *Wisconsin* and *Missouri*, 51 had hit their targets. Ninety-five percent? Terrific. The first in-flight pictures of "smart" bombs showed one going right through the door of a bunker; another plunged down the ventilation shaft of its target building.

Around the world, coalition partners were pleased with the initial progress of the air phase and people began to feel that the war might be completed in just a matter of days. There was almost a feeling of euphoria, which that was exactly what world leaders and the coalition did not want. If the public became too convinced that it was going to be a short war and the situation changed to require an extended effort, morale and support would plummet.

Everyone from President Bush to Dick Cheney to General Powell cautioned that it was too early to see the end. Brigadier General Neal repeatedly followed good news of successful strikes with cautions against overestimating the importance of such success.

Some extreme advocates of air power were claiming that aircraft alone could force Saddam to his knees.

Bovine scatology.

There was never any doubt in General Schwarzkopf's mind that when the air forces had done their thing, the mudsoldier would have to slug his way into Iraq and take personal charge of administering the coup de grace. Only ground forces could ferret out surviving enemy forces (many of them in all probability still eager and capable of significant resistance), and force them into submission. Only ground forces could receive and assemble POWs into secure camps, thereby preventing the resumption of fighting.

Schwarzkopf knew how his forces were taking the outbreak of hostilities. He wondered how the American public was taking it. Except for times when possible intelligence information might be coming in, the TV in the war room was shut off. There were some broadcasts from Baghdad, normally by CNN, although these reports were suspect.

Schwarzkopf might be curious but he had no need to be concerned. The public was gearing up for the biggest display of military support since World War II. In America, shops were selling out of yellow ribbon. Merchants couldn't keep flags in stock and they were flying everywhere, a number of them lighted and flown around the clock. An article in *Time* magazine (January 28, 1991) reported that a disabled Vietnam veteran paid 45 dollars to have an Iraqi flag made just so he could burn it.

The protest movement was alive but getting little support from the general public. A group of college students, joined by several other citizens, staged a sit-in on the median of Nevada Avenue in Colorado Springs and vowed to remain there until the war was over. They displayed the usual signs of disagreement with the U.S. decision to use military force. The city tolerantly ignored them, and a dry-cleaning establishment countered by offering to clean any American flags free of charge.

Everyone was talking about the war, of course, and the ratings of TV prime-time entertainment plummeted as viewers remained with stations carrying war news. All over the U. S., crime rates dropped—but quickly recovered. Church attendance rose and the services almost always included prayers for peace and the "safety of our men and women in the Persian Gulf."

There was some talk of fuel conservation but there

was no concentrated effort as there was still a plentiful supply, although the prices shot up with the first explosion in Baghdad. Some folks worried aloud about the fate of women who might be caught in combat; all thought it inevitable. And for the first time, fathers and mothers found themselves shipped off together as grandparents or other relatives stepped in to take over.

Photographs of Suddam Hussein became very popular—on dartboards. And that special bit of Americana, customized and printed T-shirts, flooded every mall in the nation, many emblazoned with DESERT STORM emblems and wording proclaiming each one to be an "official DESERT STORM T-shirt."

Airports, government buildings, and military installations beefed up security against possible terrorist attacks, although the main threat seemed to be sprayed-on graffiti by antiwar vandals. Coalition countries found themselves acting in similar fashion, the degree of support or number of protests varying with the internal politics of the individual country. Some Brits cheered when the B-52s took off from their bases in England while others demonstrated against British participation with the same fervor of their brethren in the colonies. In San Francisco, protesters closed the Bay Bridge—but in San Francisco, someone was always closing something. Even most of those who disagreed with Bush's decision expressed some sympathy for the difficulty he must have had in making it.

Military enlistment rates faltered only a percentage point or two. Some young men spoke of going to Canada if the draft were to be reinstated, but only a very few, and some of them were most likely caught up in the moment. Besides, Canada no longer harbors draft dodgers—it extradites them.

The Vietnam veterans seemed to be split in their support but even the split was one of philosophical differences and dread that another Vietnam was in the making.

Everyone was heartened by the success of coalition air forces and the low casualties, although the treatment of POWs in Hanoi were memories that quickly resurfaced. The Hanoi Hilton had long been shut down but people feared there would be a grand opening of a "sister" hotel, the Baghdad Biltmore in Iraq.

Now that the die had been cast, what kind of postwar peace could be expected? Surely, it was too early for those questions. But they wouldn't go away and it was a legitimate concern.

Schwarzkopf spent his days and nights within the CINC's war room and his small personal quarters. He could not allow himself to be overly concerned with the outside world. When he finally did get a few moments to pull back the camouflage cover of his bed and rest, he could comfort himself with the thought that he was doing his best. He could lull himself to sleep perhaps by recalling one of his troopers singing "Danny Boy" during one of his early field trips, or taking boyish pride in his status as an honorary Chief of the Osage Indians—he had to fight well—more often, his last conscious thoughts would be of the face of his wife and the smiles of his daughters and son.

Sleep, sleep, sleep for a while, the gods of war would whisper. Sleep well, for soon you must resume the urgent business at hand.

CHAPTER 7

THE HIGHLIGHTS (though lowlights would be more descriptive) of Monday, January 21, were that Saddam Hussein lobbed ten Scuds toward Saudi Arabia and heightened his propaganda war by borrowing an old tactic from the North Vietnamese: the television display of coalition POWs.

There is a special terror when one is awakened by the wail of air raid sirens; it is the sleepy terror of the uncertain. What is coming? Bombers? Missiles? Where will the weapons fall? On military targets? On civilians? How long will the attack last? Where will I be safe?

If the attack was by Scuds, would they have conventional or chemical/biological warheads?

Although Israeli cities went on alert, the missiles of Saddam that sped through the Monday morning skies were aimed at targets in Saudi Arabia. The CNN duty correspondent at Dhahran, Charles Jaco, had genuine fear in his eyes as he remained outside the air raid shelter and sought to narrate the attack. It was on again and off again with his mask, and time after time he announced that everyone was being ordered into the shelters. But he defied the order until the very last minute. To his credit, he gave the viewers a taste of that early morning terror.

The lights of the major city behind him were still all lit. There was actually no danger of an air attack and the Scuds, already predestined for their targets, didn't care whether the lights were on or not.

Suddenly, there was a bright pinpoint of light on the city horizon and a silver streak leaped skyward. Then another. The Patriots were on their way. One disappeared into the low cloud cover, then the other. A descending fiery ball could be seen directly in front of the fast-climbing defense missiles. Within a second or two, there appeared the dull orange glow of an explosion as Patriot and Scud met head-on and fiery debris streaked toward the ground. There were undoubtedly unheard cheers from the Patriot crews and those at risk.

On this particular night, nine out of the ten Scuds were successfully intercepted. The tenth fell harmlessly into the waters of the gulf, the victim of bad aim.

The Patriot was rapidly becoming the high-tech hero of the war: First time up and it batted .1000, assuming none were fired at the misaimed Scud. On the scene for over a decade, the weapon system had been steadily improved from its earliest antiaircraft capability to its present day function as an antimissile system. Almost

General H. Norman Schwarzkopf.

Schwarzkopf's father, Colonel H. Norman Schwarzkopf, head of the New Jersey State Police *(2nd from left)*, escorts Charles Lindbergh during the 1934 Lindbergh Kidnap trial. *(Reuters/Bettmann)*

Front row center, Cadet Schwarzkopf was a skilled member of the Valley Forge Military Academy Debate Team.

Schwarzkopf as a tackle for the Valley Forge Military Academy football team.

Cadet Schwarzkopf.
(U.S. Army Photo, U.S. Military Academy, West Point, NY)

Completing a shot put at West Point.

Major Schwarzkopf and a Vietnamese paratrooper help a wounded comrade to safety after a Viet Cong mortar attack at Duc Co. *(AP/Wide World Photos)*

A pensive Major Schwarzkopf.
Vietnam, 1965. *(Gamma/Liaison)*

During his Thanksgiving, 1990, visit to U.S. troops in Saudi Arabia, President Bush inspected the cockpit of a U.S. Marine Corps Harrier jet. *(Reuters/Bettmann)*

General Schwarzkopf addresses men of the 354th Tactical Air Force Wing during the early days of Operation Desert Shield. *(Reuters/Bettmann)*

A pensive General Schwarzkopf gazes from the window of his small jet on his way out to visit U.S. troops in the desert. *(AP/Wide World Photos)*

February 10: Defense Secretary Dick Cheney is escorted from a press conference by Chairman of the Joint Chiefs of Staff Colin Powell *(left)* and General Schwarzkopf. *(Reuters/Bettmann)*

Lieutenant General Sir Peter De La Billiere, Commander of British forces in the Middle East.
(Reuters/Bettmann)

General Roquejoffre, commander of French air and ground forces. *(Sygma)*

The proud father with one of his daughters. 1973. *(Gamma/Liaison)*

A peaceful moment with son Christian. 1966.
(Gamma/Liaison)

First Lady Barbara Bush joins in applauding Mrs. H. Norman Schwarzkopf during President Bush's State of the Union address on January 29th. *(Reuters/Bettmann)*

completely automatic, it required human input only to select a target (by priority if there are multiple targets), verify it as enemy, and make the firing decision. The battery commander receives the earliest information that an enemy launch has been made via a sophisticated space detection system. This "eye in the sky" reads the thermal signature of the launch, pinpoints its location, and predicts a general flight path.

The heart of the complex weapons system is the multifunctional phased array radar that picks up the target, locks on to track it, identifies it as friend or foe, and even has an ECCM (electronic counter-countermeasure) capability. The radar also provides guidance to the Patriot missile (which also has its own homing radar) as determined by the computers in the Battery Engagement Control Station.

There is no guarantee that the Patriot will strike its target head-on, as the interception point depends upon the two trajectories and the angle of contact. Any hit, however, downs the incoming missile, although the warhead may remain intact and explode on contact with the earth. Normally, in the wide open spaces of Saudi Arabia, the interception was accomplished away from inhabited areas. Sadly, one deadly strike was to come.

During the day, Saddam chose to display on television the captured coalition airmen, and they were presented one by one, seated or standing before a drab unicolor background. There were thirteen of them: Eight American, two Italians, one Kuwaiti, and two British. All appeared exhausted and in poor physical and mental condition. They made statements, obviously under duress; no one paid any attention to the Iraqi propaganda they were forced to voice.

U.S. Navy Lt. Jeffrey Zaun, a designated A-6 navigator-bombardier, had obviously suffered significant facial injuries, either from the effects of his ejection from his stricken aircraft or beatings by his captors or both. Prompted by the memory of mistreatment of Vietnam POWs, most viewers assumed he had been severely beaten. Only after his return was it learned that while he *was* mistreated, his face also exhibited damage from the wind impact of his 500 MPH ejection and an attempt to injure himself as a ploy against being forced to appear before the TV camera.

British Flight Lieutenant John Peters looked even worse. Unable to lift his bruised and battered head, he could only mutter in a valiant attempt to satisfy his Iraqi captors. Among the millions of viewers, there was real concern that his injuries were life-threatening.

Chief Warrant Officer Guy Hunter Jr., USMC, had a swollen and discolored left eye.

Italian pilot Captain Maurizio Cocciolone's eyes had a dazed look over his swollen nose.

General Schwarzkopf was outraged, and such pictures burned into his heart and soul. There was worldwide anger at such treatment, so flagrantly in violation of the Geneva Accords. Saddam undoubtedly thought that such a display would weaken the allied resolve. No way. We had been down that road before, too. Such brutal criminal behavior cried out for retribution. The subject of war crimes began to appear in the world press and in everyday conversation.

The mistreatment was not a surprise. It was almost

a foregone conclusion and allied airmen all knew that was it was an additional risk that they would share with the ground and sea forces. But it was a shock. And regardless of the excellent progress of the air phase of DESERT STORM, the sight of the POWs was a sobering one.

There would be more.

Shaw and Holliman of the CNN Baghdad news team had left Iraq when Saddam shut down all foreign broadcasts soon after the air war started. Peter Arnett had stayed and sixteen hours later was back on the air under the scrutiny of Baghdad censors. Arnett remained almost constantly on the tube, limited in what he could say, and although CNN anchors did repeatedly announce that their reports from Baghdad were censored, the network continued to feature Arnett in its continuous coverage of the war. It was in this situation that CNN lost some of its credibility for unbiased on-scene reporting.

By Tuesday, some 8,000 sorties had been flown. A large number of those had been support flights, such as in-air refueling, AWACS, and logistic missions. Mobile Scud launchers had become a prime target but were very difficult to find. It was felt that very few of the Scud launches were from fixed sites as these had been heavily attacked. The night air searches were intensive with the pilots using their infrared detection systems which turned the darkness into an eerie dull green picture on their scopes.

Strike aircraft continued flying around the clock and intelligence indicated that they were having considerable success against Saddam's command and control network and air defense installations. The B-52s were employing carpet bombing onto the area where the Republican Guard was deployed along the northwestern Kuwait-Iraq border but it was suspected that the elite troops were reasonably well protected by a series of underground bunkers which could remain self-sufficient for weeks at a time. Carpet bombing involved scattering bombs (normally 750 pounders) over a large area, and was particularly suitable for missions where pinpoint accuracy was not required.

The French in their Jaguars had joined the British in their Tornados in the daring displays of NATO-style low-level night attack and the long-time allies were taking out targets at an impressive rate, so much so that the British gave their Tornado pilots a brief respite by switching to the higher flying but equally aggressive Hawker Siddeley S.Mk 2B Buccaneers. The Free Kuwaitis in their outdated American-made A-4 Skyhawks were also giving an excellent account of themselves, while on their first mission the Italians were thwarted by bad weather which interfered with the necessary in-flight refueling. All but one of their aircraft on the first strike returned to base while the lone Tornado, piloted by Captain Cocciolone, continued on and was unfortunately shot down, making him the the first Italian POW.

Saddam continued his criminal behavior by announcing that he was ordering his troops to station coalition POWs at high-priority target sites. This inhumane violation of the provisions for the care of POWs reinforced world opinion that he had absolutely no regard for human life other than his own. In his concern for his own life, however, he was staying well out of the public eye and continuing to surround himself with his most loyal bodyguards.

The coalition, of course, could not let Saddam's decision affect their conduct of the air war. The last thing they wanted to do, naturally, was injure or kill their own people, but such a tactic could not be allowed to interfere with their all-important targeting objectives. There were many targets and there would not be enough POWs—yet—to allow the Iraqis to cover all of them. There was no choice but go on and pray that Saddam would not carry out his threat.

The Saudis carried out a particularly satisfying air-to-air mission during the last week in January. One of their pilots, Captain Ayed al-Shamrani, flying an American-built F-15, had succeeded in downing two Iraqi Mirages that were apparently headed for open water where British warships were operating. The Mirages were known to carry the lethal Exocet air-to-surface missiles, so effective against the British in the battle for the Falklands. Captain Ayed returned a hero, as well he should have, and it seemed appropriate that a Saudi pilot had made the crucial double kill.

Late January also saw a resumption of Scud attacks, with two aimed at Saudi targets. The reliable Patriots took out one, and the other fell harmlessly onto the desert.

But new threats were developing: not military threats, but senseless acts of violence against the environment and oil reserves of Kuwait. Oil well fires were reported by returning pilots and apparently numerous oil storage tanks had been set ablaze as well. Great columns of black smoke climbed into the air and fanned out over the battle area. Visibility at many flight levels was dropping rapidly. What was Saddam trying to accomplish now? Was it just a futile effort to make air strikes more difficult? Except for in-flight visibility there would be no appreciable military effect. However, coalition pilots were getting concerned about the hundreds of simultaneous missions within a relatively small airspace. Some were more concerned about mid-air collisions than about enemy opposition! And no one was anxious to prove the old physics axiom about the impossibility of two bodies occupying the same space! The fires would not help that situation and would certainly cause long-range damage both to the valuable oil reserves and the atmosphere over a large segment of the earth. It was another act of a madman, and a preview of the scorched-earth policy that would follow during Iraq's retreat from Kuwait.

The inevitable happened. No weapon is 100 percent effective, and the top guns of the missile world failed to intercept three Scuds that subsequently landed in Tel Aviv. Three people were killed and a number injured. At first there had been fears that the warheads carried chemical or biological weapons. The Israelis had reacted to the threat of the incoming Scuds as always, donning their masks and closing themselves into sealed rooms in the upper levels of homes and offices where they would be safe from poison gas, which tends to sink, being heavier than the atmosphere. The Scud loved multistoried buildings and tore into residential areas with an explosive fury that shattered not only bodies but the faith that the Patriots would infallibly prevent such things from happening.

Rescue teams began searching through the debris, pulling out the dead and wounded, and the ever-present TV coverage revealed the rubble and bodies and crying women and scared children.

Schwarzkopf knew the Israelis were making a valiant effort to keep their promise not to retaliate but he also knew that such a promise had a limit. It was almost reached with this latest attack. He also knew that such a thing could happen in Riyadh or some other part of Saudi Arabia, but there was little he could do about it except have faith in the Patriot system and its crews. He called for reemphasis on finding and destroying the mobile launchers, which continued to be priority targets assigned by the people in the "Black Hole."

Saddam, of course, was using the Scud in an effort to provoke Israel into a military reaction, just as he had done all along. And few would have blamed them if they had retaliated. Perhaps the only thing that saved the day was that the Scuds carried conventional warheads. The sight of gassed civilians, including children and babies, might well have been too much.

If such a thing were possible, the war was taking on an even more sinister look, with hellfires and deliberately targeted civilians.

On Thursday, January 24, Saddam cranked up his propaganda and misinformation machine and made a new outrageous announcement. Coalition aircraft had deliberately bombed a baby-milk factory. Baghdad television ran pictures of the ruined "plant" every few minutes. The rest of the world knew the claim was ridiculous. Even if it *had* been a milk production facility, it would not have been "deliberately" targeted. But there was "undeniable" evidence—at least from the Iraqi point of view. A hand-lettered sign stood in front of the rubble, saying in English: *Baby Milk Factory.* English? Yep. Right there in five-inch-high letters above the Iraqi language description, marked plainly for any attacking pilot to see! Iraqi TV made certain that a number of mothers holding infants were among the crowd viewing the ruined building.

As for the real nature of the site, intelligence sources had confirmed it as a biological production facility.

On January 26, nightime brought more Scuds, aimed at both Israel and Saudi Arabia. But the Patriots vindicated themselves and all were intercepted and destroyed.

Coalition air strikes continued.

Back in the States, a few people were saying "nuke 'em" without realizing the horrible results of such action. To begin with, the coalition forces had no intentions of using nuclear weapons, although there were undoubtedly such devices on the big aircraft carriers. They were part of even the peacetime complement of weapons on major U.S. Navy warships, always at the ready, but even aboard ship so closely guarded and securely stowed that there was no possibility of their ever being used without authority.

Nukes were not needed. Everyone knew—except possibly Saddam—that the provisions of the U. N. resolutions, as supported by the forces of DESERT STORM, could be accomplished by conventional means. Nuclear attack would be overkill of the highest order—and the worst kind. But the mere existence of such weapons of mass destruction was a reminder that Saddam Hussein had his own nuclear development program going. The general consensus was that if he had a nuke he would use it. Such was the killer instinct within the man. The Israelis had put a crimp in his program years before by their preemptive air attack on his production facility, and now coalition aircraft had sought out and attacked his current plant.

Those who advocated U.S. use of nuclear weapons were running their mouths before engaging their brains. World opinion would certainly not support such a tactic, including the American people.

Nukes were a last resort in a doomsday scenario and as much as Saddam might have liked it, he just couldn't create that kind of threat.

The world is whole today because the nations that do have nuclear weapons have the maturity and wisdom *not* to use them. Their real value is as a deterrent. Over forty years of Cold War confrontation has been witness to that fact. If during the worst of those years, the United States and the Soviet Union had not had those weapons, there would probably have been a catastrophic conventional war. The tenets of Marxist-Leninist Communism would have demanded it and there would not have been a sufficient deterrent with conventional weapons.

There are great numbers of people who do not like nukes. In fact, no one in his right mind *likes* nukes. But they are a fact of life and the world's real problem is keeping them out of the hands of the Saddam Husseins. That is getting more difficult every day.

There were those who thought Saddam should be assassinated, and strong arguments could be made for such an act. But political assassination, even in a wartime environment, brings lots of questions to the fore. Certainly, it would have been fortuitous if some Iraqi had decided that enough was enough. And the period of world mourning would have been measured in nanoseconds.

Even President Bush, who must be extremely careful in addressing such a sensitive issue, declared that Saddam should be brought to justice, an event that would certainly cause him to be removed from office at the least. There would be no assurance that a succeeding government (or ruler) would be an improvement.

So the war entered its second week with spectacular accomplishments, remarkably low casualties, and renewed confidence within the coalition that they had been melded into a precedent-setting team.

Such an accomplishment would not have been possible without each of the member nations deferring to the others on some matters. It was a diplomatic as well as a military achievement and a large part of the credit had to go to that hulking bear of a man cloistered in his below-ground headquarters and getting madder every day at the cruelty of his number-one opponent.

CHAPTER 8

ON FRIDAY, JANUARY 25, the number of sorties flown passed the 15,000 mark. One of history's largest air offensives was laying ruin to the military assets of Saddam Hussein, both in Kuwait and Iraq. Coalition pilots were demonstrating that all of the monies poured into their modern aircraft and the computer-dominated technology that enabled them to fly and fight were well spent after all, primarily because everything worked as advertised. It was an expensive phase with a push of the firing button easily sending off a half-million dollars' worth of weapons.

But Schwarzkopf's super-aggressive campaign would save coalition lives, and in the long run Iraqi lives as well.

As the weekend passed, the burning oil facilities were

sending out even more smoke and there were reports of oil slicks drifting south along the northern Kuwaiti shoreline. Saddam claimed they were from installations damaged by coalition attacks. The coalition claimed they were deliberately created by Iraqi forces. Before the conflict had started there had been reports that one of Saddam's tactics was the creation of a huge oil slick off the beaches of Kuwait. It would be lit to foil any amphibious assault, a bizarre plan and not considered by many to be a viable course of action. Floating crude would be very difficult to light, although it could be done. It was known that the Iraqis had dug trenches along their defensive lines opposite Saudi Arabia and had indicated that they were filled with oil which could be ignited to form a deadly fire barrier to advancing coalition forces.

What other "tricks" did the madman have?

One of the big dangers of the oil leak was that gulf currents would take it south along the coast and it would eventually endanger the Saudi port of Jubail. There, the Saudis had constructed one of their largest desalination plants and there was no way the facility could purify the water if it ingested oil. If the seepage were not stopped soon that would be a very real possibility. And if it reached Jubail and the desalination plant was shut down, 80 percent of Riyadh would be without water; and if it reached the desalination plant at Azziziyah, fifty miles south of Jubail, the entire eastern province of Saudi Arabia would be threatened.

The location of the leak was a pumping station at the Kuwaiti port of Al-Ahmadi, just south of Kuwait City. Kuwaiti authorities agreed that something should be done, but they had reservations about destroying the facility. There was already a 600-square-mile slick heading south. It was a much larger spill than that from the *Exxon Valdez*, probably on an order of ten times the size. It would be possible to hold off some oil around Jubail by floating large-diameter plastic pipe around the area to be protected. The oil layer, being only inches thick, would be held back by the pipe (called a boom), and an attempt could be made to suck up some of the oil by skimmer boats. Some oil would evaporate and eventually break up into tiny globules. But by that time, marine and shore life would have suffered; environmentally, it was disaster on a grand scale.

As bad as the situation was, Schwarzkopf could only consider the effect it would have on military operations, and that would be minimal. The Navy claimed it could still put marines ashore if required.

Of more immediate concern were the incoming Scuds that arrived once more in the darkness of night, both in Israel and in Saudi. Two deaths were added to the list of terror kills and the coalition continued its efforts to wipe out all of the mobile launchers in Iraq, a difficult task.

On Sunday, January 27, another unforeseen event took place. Iraqi military aircraft were flying to Iran, some of them the MiG-29s. Upon landing, they were impounded and Iran declared they would not be released for the duration of the conflict. There was speculation that the pilots had defected. Although Iran insisted the planes would not be allowed to take off, they still had to be considered a threat; in Iran, the aircraft would be in better position to attack naval targets in the gulf than they had been in Iraq. But it was doubtful that Iran would permit such a thing, as it would draw the country into the war.

There was talk about attempting to intercept them, but at the moment they presented no threat to coalition forces and air resources were needed elsewhere. Instead, further movement would be monitored by the airborne AWACs and the option to intercept left open.

Within the next days, a few loyal Iraqis rose to test their skills against the allies, and coalition F-15s promptly shot them from the skies (three kills). But there were U.S. casualties in other areas. A Special Forces AC-130 Pave Spectre gunship, a heavily armed version of the familiar C-130 Hercules cargo/troop aerial transport, was shot down while trying to knock out an enemy ground position. Not a widely publicized craft, the AC-130 carried a deadly array of 20-mm gatling cannons and 40-mm twin Bofors cannons, all firing from the left side while the pilot kept the gunship in a left-hand circle. Detection of ground targets was by infrared sensor, low-light television, and other devices that gave it a special night capability.

It fell offshore and all fourteen crewmen were lost.

Two American soldiers, stranded with a disabled truck on a supply mission to the forward lines, disappeared and were presumed captured, fallen into the hands of the Iraqis. One was female. The woman, Army Specialist Melissa Rathbun-Nealy, won the dubious honor of being the first American female POW of the conflict.

While 75,000 protesters marched in Washington, D. C., the U.S. submarine *Louisville* launched the first cruise missile ever shot in combat from that type warship, a Tomahawk that reportedly flew inland and unerringly hit its target. The Navy, meanwhile, was scouring the waters of the gulf for remnants of Saddam's small navy. Fifty vessels had already been destroyed.

And Patriots continued to knock down Scuds.

It was Monday in Kuwait when coalition attack aircraft hit the source of the Persian Gulf oil leak and seepage slowed to a trickle.

Back in the States, it was Superbowl Sunday and Whitney Houston belted out a rendition of "The Star Spangled Banner" that had the crowd cheering and waving small American flags like never before. There had been some talk about cancelling the game due to possible terrorist attacks, but the traditional end to the football season was too much a part of the American scene to drop. It was decided that the game would be played as scheduled, but under the most rigid security precautions ever instituted at an American sporting event.

A day later, Iraq played one of its propaganda cards, announcing that captured allied pilots had been injured in coalition air raids. The raids continued.

Within the CINC's war room, an attempt was being made to evaluate the damage done to the Republican Guard by the massive B-52 carpet-bombing raids and attacks by other coalition aircraft. A great number of the Guards were suspected of being hunkered down inside elaborate reinforced bunkers that had been constructed under the desert sands. Intelligence sources indicated that as many as a thousand troops could be housed in a series of interconnected concrete cylinders, each of which provided living quarters for perhaps a hundred troops. The underground complex was complete with mess facilities, sick quarters, and ventilation, protected by a "roof" of reinforced concrete several feet thick, with desert sand over that. The shelters were even reported to be effective against nuclear attack. If the reports of the bunkers were accurate and the Guards had access to them, they could emerge for the land battle practically unscathed. It was a very serious question mark in the deliberations on how long to continue the air phase of DESERT STORM.

There had been ground skirmishes along the border, mostly minor artillery duels, but on Wednesday, January 30, a major exchange occurred between Iraqi guns and U.S. Marines as the leathernecks attacked an Iraqi bunker just across the border in Kuwait. There were no U.S. casualties, although some damage was inflicted upon the Iraqi position. As the week ended, however, a much more serious engagement took place.

Perhaps to test the strength of allied forces, a column of Iraqi troops and armor crossed over and into the Saudi city of Khafji while a smaller contingent approached Umm Hujul, some fifty miles away. Khafji was a deserted place, the occupants gone for some time to avoid just such a threat, and the Iraqis occupied it with no resistance.

The next day, however, Saudi and Qatari forces launched a ferocious counterattack that culminated in street fighting and the taking of five hundred Iraqi prisoners. The coalition troops were supported by U.S. Marine artillery and air attack. The Apaches and

Cobras, and the ugly A-10 Warthog ground-support aircraft lived up to their advance notices by taking out almost all of the Iraqi armor along the border, as well as at Khafji and Umm Hujul. Tragically, eleven marines were lost to "friendly fire." Early reports indicated that one of the A-10s had mistakenly identified a marine armored personnel carrier as enemy. There were similar enemy vehicles.

The terrible incident generated considerable discussion at home and among the media, while the military immediately started an investigation. At the daily Riyadh briefings, Brigadier General Neal was pressed by correspondents to explain how such a thing could happen. It was a legitimate question, but it was asked by people who had no experience with the complexities of vicious combat between forces that are in close contact with one another. It was a case of mistaken identity, pure and simple.

How could that be? asked the media. Another legitimate question, but one that had an old and familiar answer. The pilot of an attack aircraft, particularly one like the extremely low-level and highly maneuverable A-10, must do a number of things all at once. First of all, he must control his aircraft, no small feat as he pulls maximum G-loads in ninety-degree banks—or more—just a few feet off the ground. One false move and he goes smashing into the sand at several hundred miles per hour. He is "jinking" frantically to prevent the enemy gunners from bringing their weapons to bear on him and his head moves constantly as he continually checks his attitude, flight instruments, enemy reactions, and the numerous switches necessary to select, arm, and fire his preferred weapon—missile, bomb, or gun. He is literally doing a dozen things at once, all of them critical, and any one of them fatal if he does them incorrectly. But that is what he is trained for. Nevertheless, add the tension of being under fire and an anxiety to kill the enemy, and you have taken him out of the training environment. As his aircraft speeds up and down and around, he has only an instant to recognize a target (usually by eyeball first and sometimes by eyeball alone) and fire his weapon. Just a split second to detect, recognize, and fire. A good cough takes longer.

Usually, a pilot does not know he has killed his own troops until he returns to his base. It is a shock that universally brings tears and severe remorse, and it can affect him in such a way that he may never fly again, even though he knows that "such things happen." It is not the same when it happens to *you*.

It occurs in every war. A soldier doesn't hear the password and kills a member of his own unit. An artillery battery isn't aware that friendly troops have just stormed an enemy position that is its next target. An antiaircraft battery hastily fires at an approaching aircraft. A Warthog sends a missile into a friendly personnel carrier.

When friendly fire does result in casualties, an investigation, as mentioned above, is undertaken to determine the culpability, if any, and what steps can be taken to see that it does not occur again. In the case of the above incident, all mobile ground units were ordered to place a special identifying mark on their vehicles that allied pilots could readily see.

The investigation revealed that an air-to-ground missile was responsible, but the pilot was assigned no blame.

The battle for Khafji resulted in complete defeat for the Iraqis, though they had fought well and inflicted serious casualties on the Saudi and Qatari forces. There were reports that as many as eighty percent of the Iraqi armor had been destroyed.

One Iraqi tank column had crossed the border with its guns pointing toward the rear, an indication that it wanted to surrender. When approached, however, the guns swiftly came about and began firing. Was it a deliberate subterfuge? Quite possibly, and allied troops were warned to be alert for further deception.

Schwarzkopf and his staff were constantly fighting the battle of damage assessment. They had photos and pilot reports and space reconnaissance but things were not always black and white. Bridges had been knocked out, but they could be bypassed temporarily by pontoon-supported roadways. Airfields had been severely cratered, but holes could be filled in and patched over. Troop assemblies had been attacked, but there was no way to tell the number of dead and wounded. Armor was a bit different. A destroyed tank was too heavy and difficult to move, as were mobile artillery units and other vehicles: Some count could be made of such targets.

Mobile Scud launchers, very high priority targets, were apparently being destroyed, and the number of Scud launches was going down almost daily.

Coalition air losses were very low, so much so that on Wednesday January 31, Schwarzkopf announced with confidence that air supremacy had been achieved. Sorties could proceed almost at will at Iraqi targets.

It would be interesting to know why the Iraqi Air Force put up such weak resistance. It was almost nonexistent. The MiG-29 was a state-of-the-art fighter, and Iraqi pilots undoubtedly received Soviet training. But their reluctance to fight gave free rein to allied air attacks on critical ground targets. Could it have been a matter of command and control? That was probably a factor, along with the Iraqi inability to coordinate fighter attacks against enemy aircraft. With modern high speed jets, some form of ground or air vectoring is required to get defensive forces in position to engage the attackers. Against Iran with comparable technology, it had been different; just take off, fly to Iran, and attack. It is probable that the sophisticated systems necessary for Iraqi airmen to engage the up-to-the-minute coalition forces just did not exist. They knew that if they took to the air, they would be shot down.

In any event, coalition reports will someday reveal the reason.

January was rapidly drawing to a close. The air phase had achieved many of its objectives but there were a considerable number of tasks left. The Republican Guards would be continually pounded. Communication and command and control facilities would be subject to reattack. It was essential to separate Iraqi ground forces from their commanders: Cut off the snake's head and it may wiggle for a while, but it has no sense of direction or purpose.

In early February, the battleship *Missouri* unleashed its big sixteen-inch guns against Iraqi targets in

Kuwait, perhaps in rehearsal for the naval gunfire it would be required to supply when the ground offensive got under way. It was the first time in forty years the "Mighty Mo" had fired in anger.

By the end of the first week of February, the Syrians had joined their Arab brothers in combat, repulsing an Iraqi border probe with a minimum of casualties. Schwarzkopf watched the fighting performance of his Arab forces with considerable satisfaction. When the ground war came, they would play an important part in critical tactical maneuvers. Thus far, they had bested their opponent by a significant margin. The coalition was proving itself.

The Iraqi pilots trying to steal away to Iran finally got their comeuppance when U.S. F-15 Eagles shot down four of them. The "Iraqi detachment" in Iran had reached more than one hundred planes. Iran reiterated that pilots and planes were interned for the duration of hostilities, although she was becoming a bit more sympathetic to the Iraqis.

Also at the end of February's first week, the battleship *Wisconsin* had ended its own forty-year famine and lobbed over one hundred 2000-lb shells onto Iraqi targets in Kuwait.

At the start of the second week in February, SecDef Cheney and General Powell left Washington for the long haul to Saudi Arabia. Their purpose was obvious: to get updated in person by Schwarzkopf, and discuss with him an appropriate time to start the ground offensive. With their departure, a flurry of media speculation began. The media military consultants were breaking out fresh charts and detailing how they "would do it."

Such activity didn't set well with Schwarzkopf. Not at all. Here were professional military men who had advanced to senior positions within the military discussing his (Schwarzkopf's) options in conducting the ground war. Saddam Hussein had excellent television reception and the experts were discussing established tactical considerations on open networks. There was no question about their integrity; they were just speculating. But speculating from their level of expertise, many of their arguments were quite valid. Whether by design or accident, several of them discussed tactics that Schwarzkopf had under consideration. Sure, some schemes of operation would be obvious to a first-year military academy student, but others were too close to home. Almost all mentioned a flanking move, something Schwarzkopf had gone to great length to set up as an option, with his rehearsal of amphibious landings aimed toward drawing Iraqi troops east toward the coast. Now that he had taken out most of the intelligence-gathering ability of the Iraqi generals, he was starting to shift the main bodies of his troops *west*, to get them into position for his "Hail Mary" flanking attack. This was *his deception plan!* And American television was openly discussing it. By accident, of course, but by qualified analysts.

What were the networks trying to do? Most of their viewers had no detailed military background. Why go into elaborate explanations of military "options" that were most probably a bit boring to many in the prime-time audience? It did no good to shift channels; everyone was doing the same thing. It seemed that television anchors were vying for the best analysis of what was to come. Maybe one would even be right on and Schwarzkopf's whole campaign would be revealed! That could be good for an Emmy or two ("We showed it first!").

Obviously, the objective was ratings. Throughout the war, trade periodicals had been discussing and rating war news broadcasts: A matter of good business, perhaps, but was any serious thought given to the fact that, by presenting bona fide military experts, some valuable information could be presented to the enemy? Maybe just a piece. Maybe just *the* piece Saddam was looking for?

Actually, the public was saturated with war news and it was difficult for television correspondents to come up with anything "fresh" under the military restrictions imposed on correspondents in Saudi Arabia.

In any event, Schwarzkopf didn't like it at all.

The Soviet Union, beset by internal problems, was attempting to gain some international prestige by supporting various peace (i.e. cease-fire) proposals cropping up within the U.N. community. For a while, it appeared that the U.S. was looking on one of these with some sympathy, but administration officials quickly put the squash on that when it was learned that the proposal also considered the Palestinian question. The United States continued its hard-line stance that any cease-fire would be predicated upon only one condition, the immediate and complete withdrawal of Iraq from Kuwait.

Secretary Cheney and JCS Chairman Powell arrived and met with Schwarzkopf and key members of his staff. While the subject and content of the discussions were not announced, it can be assumed that the two visitors from Washington received a thorough briefing from Schwarzkopf and an estimate of the situation relative to the start of the ground war. From the very beginning, Cheney and Powell had left the conduct of the war to the man in charge, and he, in turn, had kept them apprised of the progress of the DESERT STORM air phase. But on-site discussions give the top military and civilian commanders a chance to discuss details and weigh options with a more personal "feel" than one gets from a distant voice on the faraway end of a phone line.

The three top managers of DESERT STORM (Cheney, Powell, and Schwarzkopf) also met with Saudi authorities, among them Utman al-Humaide, the Saudi Assistant Defense Minister, and discussed the timing of the ground phase.

Cheney and Powell would be briefing the president on their return. It was imperative that Bush be aware of the exact situation. While Bush would approve the decision and date to start the ground war, there was little doubt that he would go along with Schwarzkopf's recommendation. But the president was involved in several delicate political discussions, and needed the battlefield plan updated for background reference in his continuing rounds of political talks.

Diplomatically, things were going well. The coalition was tight and functioning well. Both England and Spain had given permission for B-52 raids to originate in their respective countries. France had granted overfly rights—as long as the bombers carried no nuclear weapons. No problem.

Mikhail Gorbachev, perhaps smelling the sweet scent

of becoming a peacemaker, announced that coalition military efforts were going beyond the original intent of the U.N. sanctions, and sent a representative to Baghdad. If he could come up with something that had a chance to end the war, his support within the Soviet Union would grow, and he needed that. Washington watched with a wary eye.

There was always the chance that some unforeseen development, either military or political, could complicate the U.N. approval of continuing with DESERT STORM. Certainly, Jordan was becoming more vocal in condemning the air strikes that were killing "thousands" of civilians in Iraq, including Jordanians traveling into the country to deliver "non-embargo" supplies. King Hussein was at this point openly siding with Iraq.

With the visit of Cheney and Powell, speculation on just when the ground war would begin skyrocketed. It became a major part of almost every newscast. Pros and cons were discussed. An early start would most probably "get it over with" but could result in higher casualties. A delay could find public support eroding but make maximum use of the air phase to soften Iraqi resistance. However, after a while, air attacks could be just "shifting rubble" and not be cost-effective.

As for Saddam, he was taking full advantage of the situation. The longer the coalition delayed its ground attack, the more his stock rose among the Arabs who felt he was making a heroic stand against the world's mightiest power and a mighty array of allies. And if they chose to attack soon, he was ready! Boldly, he announced over Baghdad radio, "The number of Americans killed will exceed tens of thousands if a ground war occurs with Iraqi forces." A number of people feared his prediction could be close to the truth.

But Norman Schwarzkopf and his staff were in no hurry. When the time was right, the time was right, and the determination of that time would depend upon a number of things including an experienced warrior's gut feeling. Meanwhile, the air pounding would continue. There certainly was no erosion of coalition public support, even in America where protesters kept shouting their opposition but to fewer and fewer sympathizers.

And Neal at his daily briefings in Riyadh gave absolutely no indication of any decision despite constant questioning by correspondents anxious to get even a hint. Some analysts covered all the bases by saying that the ground phase was imminent and could come within the next days or weeks.

Probably so.

CHAPTER 9

RELATIONS BETWEEN the military and the media were deteriorating with respect to pool coverage, censorship, and delays in approving copy.

Pool members felt that they were not always taken to the right places at the right times. Correspondents were concerned that military censors were not only removing material from their reports that could be harmful to the war effort, but anything that could be detrimental to the military. Too many times, approved reports were released too late for timely use.

The military, on the other hand, was not able to foresee where and when top news items would be generated. Later it became apparent that protecting the good name of the military was not one of the purposes of censorship, and that practically all of the items censored from reports had been battle sensitive. As for delays in processing reports, there were thousands of reports and the physical task took time.

A number of correspondents disregarded restrictions and struck out on their own. Among the early ones were CBS's Bob Simon and his three-man team who early in the air war had donned military-type gear and taken off in a vehicle to get independent front line coverage. (They were released along with other POWs after the cease-fire.)

Some foreign correspondents had been more fortunate and scored scoops. An Associated Press correspondent had actually sneaked into Khafji and reported from the "inside" as coalition troops retook the city. French and British TV crews were in the city well before pool reporters.

Some independent correspondents, not having names or credentials with clout, missed out on the pools, and were relegated to putting together news stories from briefings and what they could pick up on TV.

Four Iraqis surrendered to a small group of correspondents who were out on their own near the border.

The military had little sympathy for the media violators. One unauthorized Associated Press team arrived at the location of the 24th Mechanized Infantry Division and was unceremoniously sent back to Dhahran.

News gathering is a highly competitive business and the reporter's fame and fortune depend on his/her outperforming the competition. It is natural that any type of restriction affecting that ability is deeply resented. Yet the majority of correspondents in Riyadh went along with requirements. Those who did not were sometimes resented by pool reporters who were "playing the game."

Schwarzkopf was very conscious of the need and rights of the public to follow the progress of the war, and as mentioned earlier, balanced that need and right against his obligations to safeguard the lives of coalition forces, to prevent any premature revelation of tactical battle plans, and to insure that the work of correspondents did not interfere with operations in the field. Obviously, releasing almost 1,200 correspondents to wander about on their own would present a certain amount of chaos.

A field commander, fighting for his life and the lives of his men, is no more enthusiastic about an on-the-spot interview than is the mother who has just seen her child run over, responding to a microphone or TV camera stuck in her face, and the insensitive question, "How do you feel about losing your child, Mrs. . . ?" The days of Ernie Pyle, when a correspondent was more interested in reporting in a more responsible fashion the horror of war, seemed to be gone.

CNN's Peter Arnett was continuing to broadcast from Baghdad. At first, this did not trouble Schwarzkopf, but as reports became more and more concerned with civilian casualties, the general resented it. One of his top priorities—and he had passed the order down the line to where it was in the thoughts of every pilot as he climbed into a cockpit—was that air attacks would be conducted with the welfare of civilians always in mind. But as Arnett reported more and more on civilian casualties, he seemed to be caught up in Saddam's propaganda machine, and instead of the more accurate "Baghdad authorities reported x number of civilian casualties," Arnett was saying "x number of civilians were killed today by air attacks." The difference is that the former statement is based on second-hand information and is questionable; the latter presents the information as factual. The "baby-milk" factory incident, which was so clearly a product of Iraqi propaganda, had received extensive coverage. Yet time after time it was reported that the people of Baghdad were going about their business in a normal fashion—as if they were confident that coalition air attacks would not disrupt their daily routine. There was an obvious contradiction.

Peter Arnett is a highly skilled correspondent, with impressive credentials including a Pulitzer Prize, who feels a great obligation to his profession; and it is doubtful that he would intentionally do anything that would endanger coalition forces or give comfort to Saddam. But a number of people felt he and CNN were being used, and wondered if keeping the window to Baghdad open was worth the price.

On Monday, February 11, Saddam Hussein addressed his people for the first time since the start of DESERT STORM. He promised them victory with his usual exaggerated expressions of Iraqi military capabilities and references to the battle experience of his troops. After all, they had fought for eight glorious years against the powerful Iranians. The coalition took his remarks in stride. They were made for internal consumption. If his address did anything significant, it showed that he had not yet been a casualty of the air attacks. Nor should he have been: He had a heavily protected bunker a few miles out of the city, and there were reports that he and his bodyguard entourage sought out exclusive residential neighborhoods when they were in Baghdad, secure in their knowledge that coalition forces would not attack such areas. He was not individually targeted although there was some hope among coalition pilots that he would be in the wrong place at the wrong time.

The air pounding of Iraqi troops continued with no letup, and it was difficult to understand how any field army could survive such punishment. These attacks made it doubtful that Saddam could maintain any instantaneous command and control links with his forces in southern Iraq and Kuwait. And that was Schwarzkopf's intent. Isolate the troops from their Baghdad authority, cut their supply lines and make them miserable in the field, inflict casualties around the clock to destroy morale and reduce their fighting capability, eliminate their armor as much as possible, and wear them down to the point where they would have little will to resist attacking ground forces.

Schwarzkopf continued to position his troops for the assault into Iraq and Kuwait. When the time came, he intended to launch a coordinated air-land attack that would feature three phases: An early attack by marines on the eastern flank into Kuwait to reinforce the Iraqis' thinking that Kuwait and Kuwait City would be the first objective; a massive incursion through breaks in the central defensive lines across the Saudi-Iraq border; and a race by coalition troops around the left flank into southern Iraq, which would seal off any avenue of escape. Surrounded, the Iraqis would have only two options: Surrender or be killed.

Saddam was no big mystery to Norman Schwarzkopf. He had studied the dictator and reviewed Iraqi actions in their war against Iran. To him, Saddam was just a man in a uniform with no military talent—and he was predictable. True, such unconventional acts as setting the oil facilities on fire and opening oil valves to the sea were shockers, but they were not surprises. Saddam's big threat was that he was irrational and possibly unbalanced. There were reports that he had been under psychiatric care at one time. In the past, he had shown no hesitancy in ordering his men to fight to their deaths, or even having them killed if they refused. He seemed to have absolutely no regard for his civilian population, placing military assets in the midst of them in an attempt to forestall attacks or at least be able to show that innocents were killed as a result of coalition actions.

It is difficult to say whether Saddam was disappointed or elated at Bush's announcement on Monday, February 11, that the president was in no hurry to start the ground war. In all probability, Saddam wondered if the president was sending the message to deceive him, so that coalition troops might attack the next day and hope for surprise. Whatever Saddam was thinking, it was the product of a crafty mind that must have seen defeat coming, but was desperately searching for a course of action that would still bring him out of the conflict with an advantage. If only he could get one or two conditions imposed on the withdrawal ultimatum. He would save face and in the eyes of his Arab friends—Jordan,

Yemen, Algeria, the Palestinians—he would have stood up to the Great Satan. He had already endeared himself to the Palestinians by insisting that any negotiations include the Palestine question. Yes, he had been rebuffed, but he had scored his points. Now he needed to win at least one set out of the match.

Shortly before the middle of February, allied land, air, and sea forces engaged the Iraqis in Kuwait with a bombardment that was the largest concentrated encounter yet. It lasted only a few hours as battles go, with the Iraqis on the receiving end; but it destroyed still more Iraqi war-making hardware, generated casualties, and served as a preview of what was to come.

Valentine's Day came and troops were swamped with cards and letters of love and support. Some of the mail was delivered after the fourteenth, but the message had lost none of its luster. It is sometimes hard to equate military people with sentimental love, but they experience a unique side of that emotion. There is no camaraderie like that of men, and now women, who have gone into combat together. It is a very special bond and undoubtedly confuses wives who, while attending military reunions in later years, see their husbands fervently hug balding, pot-bellied men they have not seen or heard from for a quarter century or more. It is probably a bit puzzling for some of the women when they find themselves alone at a table with empty drinks in their hands while their husbands laugh and joke and slap the shoulders of their war buddies. What they are seeing is a resurgence of that special kind of love one man had for another when they walked down the dark valley of death together, each depending upon the other, although perhaps moments before they had been complete strangers.

In a bizarre sort of way, those who have experienced combat even seem to have a unique love for old enemies, although that emotion is slower to manifest itself. At a recent reunion of American and Japanese war veterans on Iwo Jima, the two groups, a bit standoffish and embarrassed at first, undoubtedly harboring terrible memories of each other, spontaneously came together and embraced and cried and wished each other well. They were not only paying tribute to their lost comrades but to their former enemies who had also experienced that same sort of love. It is unexplainable to those who have not experienced it. But like all true love, it grows stronger year after year. It is a love that comes only after paying a terrible price.

The end of February arrived and most of the men and women who had been together in the desert sands, some for as long as seven months, were on the verge of entering their own dark valleys and learning about that new love.

CHAPTER 10

THE MIDDLE OF FEBRUARY provided an opportunity for Saddam to further crank up his propaganda machine. In mid-February allied aircraft attacked a bunker in a Baghdad suburb that intelligence had designated as a communications/command and control facility. The bunker had been penetrated by two "smart" bombs that had punched through the concrete of the heavily constructed roof and exploded deep inside. Saddam claimed it was a civilian shelter, and indeed it had been one when originally constructed. But allied research and observation had determined that, while it was possibly used by civilians, it did have a military purpose. The sources pointed out that the bunker had been surrounded with fencing, and had only limited access, hardly a configuration that would allow rapid entry by a mass of civilians during an air raid. And why would a civilian shelter have on its camouflaged roof communications antennae that were hardened against the EM (electromagnetic) effects of a nuclear blast? Military vehicles and personnel had been regularly recorded entering and leaving the building, but there had been few civilians.

True, casualties were in the hundreds (although not the five hundred that had been reported by Iraqi authorities the morning after) and for the next few days, the rescue efforts dominated Baghdad news. Crushed bodies were removed and one, puzzlingly enough, was carried out under a draped Iraqi flag. A VIP? There was even some speculation that it was Saddam. Whoever, the body received no special coverage beyond a passing shot of the TV camera.

To add credence to the allied claim, the basement was flooded by the Iraqis after the attack and could not be examined. The bunker had not been casually added to the target list. The targeters in the "Black Hole" had checked a number of sources as to the nature of the building, including foreign workers who had constructed the shelter. It was plain that the bunker had been singled out in the attack in such a way that the residential area surrounding it did not suffer collateral damage. It was a true surgical strike, without warning, which gave rise to the speculation that it was hit by weapons delivered from Stealth aircraft. It was.

In any event, Saddam's propaganda took hold. At the next Riyadh briefing, Brigadier General Neal was bombarded by questions from correspondents who obviously had swallowed the Iraqi line. Why are we targeting civilian facilities? How do you know the shelter was used for military purposes? One reporter even asked if coalition forces would in the future notify civilians near military targets of impending raids so they could clear the area. Neal might have been stunned by the question, but his face did not convey anything. Notify the civilians (and by association, enemy military) of targets in advance? It is doubtful that coalition pilots would have appreciated going on a raid, knowing that antiaircraft defenses could practically be pre-aimed. They were already using tactics that put them at increased risk to insure civilians were not deliberately endangered. There had been general warnings to Iraqi civilians since the war began—bombing raids alone should have sent the message: stay away from military targets. USCENTCOM briefer Neal attempted to put the matter at rest when he answered one presumptive question, "What went wrong?" with, "From the military point of view, nothing went wrong. The target was hit as designated."

The bunker was used by civilians, there was no doubt about that. But one school of allied thought was that Saddam had deliberately set the whole thing up by making the shelter available to civilians, even though the basement housed a military facility. He figured it would be attacked, and the civilian casualties would impress the world. Certainly, such a man as Saddam would have had no reservations about using his own people in a cruel propaganda ploy.

A few days later, an even larger shoe was dropped. Out of the blue, the Iraqi Revolutionary Command Council announced over Baghdad Radio that Iraq was offering to pull out of Kuwait. The Iraqis in the streets received the news as if it was the end of the war and began an instant celebration.

President Bush called it "a cruel hoax," and indeed it was. Reading the "fine print" revealed that it was pure bovine scatology—to risk overusing one of Schwarzkopf's phrases.

The complete document was nothing more or less than another attempt by Saddam to back down the world. It relayed *his* conditions, not the United Nations.

He would consider *talking* about Resolution 660 (calling for unconditional withdrawal) provided there was a cease-fire; all other U. N. resolutions against Iraq were dropped; the coalition returned to its prewar geographical positions and all weapons, including those provided Israel, were withdrawn; Israel would withdraw from all occupied territories; and several other requirements. The so-called withdrawal offer was an affront to every citizen of the world who had taken a stand against Iraq. The conditions would have seemed more appropriate if Saddam were winning the war and offering this document to his beaten opponents. A number of people, notably Lebanese, Palestinian, and Jordanian, proclaimed it to be a generous offer by a peace-seeking Saddam Hussein.

The demands, not all of them listed above, were ridiculous.

Saddam was obviously playing for time. Even a short cease-fire would give him time to assess the damage to his troops in southern Iraq and Kuwait, and rush reinforcements and supplies to the occupied area. He was hurting.

He was also desperately seeking a diplomatic solution, although a great deal of time had been provided him before hostilities began. And he was being quite arrogant in suggesting terms that were more favorable to Iraq than those offered him previously. That he could feel the coalition could be so gullible was another sign that he was out of touch with reality. Perhaps he was so insulated that he could not even form an intelligent guess as to what was going on. Often compared to Hitler, Saddam was nowhere near as intelligent as the German dictator had been. As diabolical, yes. As smart, no.

It was also possible that Saddam was trying to hold off the ground offensive under the mistaken impression that American public support would demand a pause in hostilities for at least long enough to examine the Baghdad proposal for areas of negotiation.

It was not worth Schwarzkopf's time to further evaluate the offering. It was obviously unacceptable, and other people were handling the diplomatic side of the conflict. He had to continue to prepare for the ground assault, and there was no chance the proposal would affect that. As for the reaction of coalition forces in the field, there were mixed reviews. Most expressed the opinion that it was unacceptable and it was time to continue on with the action. A few, initially unaware of the proposal's full content, briefly thought, "It could be over," but that hope quickly disappeared as the details became known.

There was an interesting aspect to the offer, however. It had come just a few days after Yevgeny Primakov,

Gorbachev's representative, had visited with Saddam, and the Revolutionary Command Council indicated the offer was made "in appreciation of the Soviet initiative." That, despite the Soviet reassurance that they were behind the U. N. resolutions. Gorbachev was playing both sides of the street and doing it poorly. Surprisingly enough, the proposal received a lot of attention and when the U. N. Security Council met in closed session some time later, there was speculation that a counter-proposal was in the making. That was not the case. The U. N. remained firm. The Soviets, seeing the flurry that the announcement had caused, began further efforts on their own without consulting any of the coalition members. It was becoming apparent that while the Soviets wanted Iraq out of Kuwait, they wanted Saddam to stay in power. He had been a staunch ally and a good customer for their military hardware. If Saddam were removed from the scene, there would be a power vacuum that could very well be filled by the West and not the Soviet Union. For all of the *glasnost* and *perestroika*, the Soviet bear was still looking for a place to relieve himself in the woods.

Bush assuredly wanted Saddam to go, but was limited in what he could do; indeed, even in what he could say. He did encourage the Iraqis to "take matters in their own hands" in the hope that there were enough responsible people around to realize the damage Saddam's policies were inflicting on their country. There was some thought of a march on to Baghdad once Kuwait was liberated, but that would require clear justification, since it would be an act that went beyond the provisions of the U.N. sanctions. If Saddam ordered his forces to use chemical or biological weapons, that might do it.

It was apparent that Saddam was responsible for much more than just an attempt to annex Kuwait. The releasing of millions of gallons of crude oil into the gulf was an environmental crime of the largest magnitude. Setting the oil fires that would be darkening Kuwaiti skies for months to come was a senseless act, and as environmentally damaging as the floating oil. There was increasing word from intelligence sources within Kuwait City that Iraqi soldiers were continuing to commit atrocities against the Kuwaitis.

Surely, there were enough reasons to brand Saddam as a war criminal, and the Nuremberg trials after World War II had set the precedent for dealing with such men. Why not a U. N. trial? Or an Arab trial by Saddam's peers? Why not put Saddam in a closed room with Schwarzkopf for twenty minutes? It is a frustrating thing for decent people when they see the havoc that someone like Saddam has wrought and realize that he stands a good chance of getting away with it. Already, there was a great deal of discussion about war reparations after Saddam was beaten. Coalition partners were promising to share the expenses of the allied military effort; that seemed to be under control. But who would reimburse Kuwait for the damages caused by the Iraqis and the damage inflicted by combat on their soil? Who would pay for the reconstruction of Kuwait City? Iraq was the culprit, and reparations would surely be a consideration of any victory. But Iraq was short of financial resources, even after plundering Kuwaiti cash and securities.

The ground war was only one week away although the date was still highly classified. And the name "ground war" was really a misnomer as the final stage of DESERT STORM would be a ground-air war where air support would be working in concert with advancing ground forces that used speed, maneuverability, surprise, and deception to surround the enemy and then destroy him. The doctrine was more formally known as the AirLand Battle Doctrine and had been officially adopted by the Pentagon in 1982. It would be a coordinated effort whereby Schwarzkopf's forces would demonstrate, for the first time in combat, the tactics developed for total assault. And for the first time, there were air vehicles specifically designed to support ground troops against armor. Before, such air support had been improvised using aircraft designed primarily for other tasks. Marine air support in World War II and Korea had been accomplished by F4U Corsair fighters which adapted well to their support role while Army troops called in whatever was available from the Air Force inventory of fighters and light bombers. There had been the Douglas A-1 during Korea and Vietnam as well, originally a Navy carrier-based attack aircraft but recruited by the Air Force as a ground support weapon because of its low-level capability and rugged construction. The Marine Corps had long established tactics for such support (a proven role of the Navy-Marine team) and the Air Force had adopted similar control procedures, so the command and control aspects were familiar. Now there was the ugly Warthog and deadly choppers like the Apache and the Supercobra that had been specifically designed for ground support and attacked with laser-directed Hellfire as well as heat-seeking and optically-guided missiles. The extremely rapid-firing Gatling gun of the Warthog sent depleted uranium rounds from its seven rotating barrels at such a rate that there was a steady stream of liquified uranium smashing through armor plate. Conventional fighters such as the F-15, F-16, F/A-18, etc. were still in the picture for ground support as all could be rigged with bombs and AGMs making them serve a dual purpose. And the Navy's rugged, snub-nosed A-7 Corsair II was still a tiger when it came to hitting ground targets, although it was being phased out of the inventory. All in all, the versatility of coalition air forces was at an all-time high. Ground troops and armor would still be required to shoulder the final dirty business of man-to-man and armor-to-armor attack but they would have the very best of aerial support.

There were specific indications that Iraqi ground forces and armor were suffering heavy attrition, so much so that Schwarzkopf could classify enemy units as heavily degraded (down to 25 percent effective), moderately degraded (50 percent effective), and lightly degraded (75 percent or more effective). His last-minute battle plan adjustments could take the location of the variously degraded units into consideration, although it was doubtful that the Iraqis could still make substantial movements. They were largely dug in and waiting.

Schwarzkopf continued moving his "Hail Mary" forces to the west: the 6th French Light Armored Division, 2nd Brigade, and the U.S. 82d Airborne farthest to the west, then the U.S. XVIII Airborne Corps consisting of the 3d Armored Cavalry Regiment, the 24th Mechanized Infantry Division, and the 101st Airborne to the east. In addition, a complete logistic support base moved west to be near the left flank forces. This remarkable movement was accomplished in a timely fashion across several hundred miles of gritty desert by thousands of trucks and other support vehicles.

The troops immediately dug in, quickly shoveling out a small trench for immediate use, then a deeper and more refined hole as the next few days passed. They were going to leave their shelters soon, but for the moment, some crazed Iraqi might come from the north and they wanted to be prepared, no matter how small the risk. As they settled down for the night, there was a great unknown out there ahead of them in the darkness. How would the Iraqis fight? How would *they* fight?

The troops settled in to wait for the order to attack. They would be moving north to reach for the Euphrates River and seal the northern escape route of the Iraqis. The soldiers knew it was getting close to time. Open fires were forbidden although small protected flames were allowed to make coffee. Mail had a hard time reaching them now, but they had rations (such as they were) and plenty of liquids. The Persian Gulf soldiers were older as a group than the American troops in 'Nam, twenty-seven being the average as opposed to Vietnam when the age was five or so years younger. Most were married and many of them, like the troops spread eastward all the way to the water, had news of new babies they had never seen. The desert was very quiet, especially at night when most incidental movement was stopped and a whisper could be heard over many yards. There weren't any terrain features to help local orientation. Everything was sand; flat sand, lumpy sand, gently rolling sand, and wet, muddy sand when it rained.

To the rear, the female support troops were pulling their own weight with no quarter given or expected. They had a minimum of privacy, mainly at latrines, but were fully integrated in their duties, living conditions, and efficiency.

Desert Shield Radio, a conglomerate of four military-run FM stations, was on the air with music and there was a scattering of small personal receivers about, but they were listened to at very low volume, preferably with tiny earphones, since sound travels so far across the sand. Most of the troops were content to be alone with their thoughts. The one vestige of privacy they had left was their memories of home and the good times in the past and those to come. Some Americans wondered how it would be when they get back. People had spat on the Vietnam veterans, they understood, but not too many had been old enough to remember. A new generation, a new war. But some word had come down through the days of preparation that the public was strongly behind them and there seemed to have been tons and tons of letters from perfect strangers before the troops moved west, a great number addressed to "Any Soldier." That was nice.

Prayer was not a rarity on those last nights before the assault. Christian, Jew, and Muslim, all praying to the same God, and if the prayers of wars past were any indicator, the men and women were not just praying for their own safety and that of their comrades; they were praying not to do the one thing that a soldier fears the most: letting down his comrades. They prayed for bravery and strength in combat, and then a million very private things. Across the border, they knew the Iraqis were also praying. Which ones would God listen to? Or did He take sides? Probably not. All were His children. An old southern song probably came to the minds of some, about the turn-of-the-century rabbit hunter who was treed by a hungry bear. He prayed like the dickens, but wasn't sure his prayer would be answered since he had not lived the best of lives. So he wound up with a very sincere plea that was part of the chorus of the song, "Lord, if you can't help me, please don't help that bear."

There was an old saying, too, that there were no atheists in foxholes. Certainly, there were not many trying to get comfortable during the cold desert nights of Saudi Arabia as the third week in February, 1991, approached.

On Sunday, February 17, U.S. attack helicopters made their first coordinated night raids on Iraqi positions. Coalition forces had as a distinct advantage over their enemy the ability to conduct at night almost the entire spectrum of military operations. Infrared and/or low-light sensors were installed in practically every armed combat vehicle, land and air. The M1A1 Abrams main battle tank had a superb night detection and aiming capability. Night goggles were available to certain ground troops, especially those in the most forward areas or engaged in reconnaissance.

Flights of British Tornados made another of their brilliantly coordinated, high speed, low-level attacks. Again, military targets were hit; but Baghdad, true to form, reported a large number of civilian casualties—130. No mention was made, of course, of the two Iraqi Scuds fired deep into Israel. They were ineffective and produced no casualties.

While the public debate over the nature of the bunker hit back on the 14th continued, Abdul Amir al-Anbari, the Iraqi ambassador to the United Nations, proclaimed that if the bombing of civilians continued, Iraq would use weapons of mass destruction—i.e. chemical and biological weapons. Such an announcement introduced nothing new to the balance of power in the gulf area. Coalition forces had operated all along under the assumption that such weapons would be used. Amir's announcement, undoubtedly made as a deterrent, had no effect on the pace or scope of operations.

SecDef Cheney and General Powell were back in the White House briefing President Bush on their visit with Schwarzkopf. Bush was extremely interested in Schwarzkopf's personal assessment of the situation as well as the impressions his two top advisers brought back with them. Numbers and statistics were of some value, but they never told the whole story, and were not exclusively the material from which decisions as important as when to start the ground assault would be made. After the meeting, Bush did linger long enough in the Rose Garden to tell reporters that he was pleased with his commanders and confident that things were progressing in a satisfactory manner. Bush was very glad that he had made the decision back in November to send sufficient forces to Saudi Arabia to provide a ground option, although in those days there had still been hope for a diplomatic settlement. His foresight was paying off. Schwarzkopf may not have had the textbook superiority for attacking dug-in and waiting defensive troops, but everyone in the planning process seemed to have confidence that the general had sufficient strength to do the job.

Part of his forces were the combat engineers. They would be the first to move forward and open pathways through the formidable "Saddam Line," the multilayered defensive barrier the Iraqis had constructed in the sand just north of the Saudi border. Waiting were thousands of land mines, some alert for the pressure of a passing vehicle or foot; other "more intelligent" devices would need only a sharp thump on the nearby sand for detonation. There was the razor wire, effective against troops but no significant threat to tracked armor. Deep trenches would have to be negotiated, some probably filled with burning oil. High sand berms had been built up. The combat engineers would have to deal with all of the barriers and they would be doing it under fire.

Aircraft had bombed some of the barrier area but most of it was still intact. Still, the engineers did have some proven clearing weapons and tactics. They had rocket-propelled hoses that could be fired forward. The long snakelike tubes were filled with explosives, which, when they detonated on contact with the surface, would clear a path forward. The similar British Giant Viper system had the ability to neutralize an area 10 yards wide by 200 yards long while the American system covered a little less ground with each "shot." Both the British and American troops who used these weapons—the Mine Clearing Line Charge, or

MICLIC—knew they were sitting ducks if enemy artillery could zero in on them. Finally, there were the foot "sappers," walking troops with metal detectors or even bayonets to locate and mark the mines for removal. This was a slow and very dangerous technique, completely unsuitable when tactics involved fast strikes into enemy territory.

The wide and deep trenches could be filled with large cylindrical bundles of chained pipes (fascines) that would be dropped in place to fill in the trench and allow armor to ride across the barrier. Other specialty vehicles could unfold steel 60-foot-long bridge sections. Armored bulldozers could break, crush, and flatten berms.

Time was truly of the essence. The longer the clearing operation took, the longer the engineers would be exposed to enemy fire. Equally important was the fact that once the engineers started clearing paths through the Iraqi defenses, fast-moving armor and troops would be pressing to plunge through the breaches. It could be a very bloody business.

As the troops waited, the air war was in its fifth week. Over 60,000 sorties had been flown. Some pilots were reduced to looking for targets of opportunity along the roads of Iraq and Kuwait. B-52s continued to fly from Britain and Spain as well as from Diego Garcia, their primary target still the Republican Guards in the southeast sector of Iraq. The venerable bomber was older than most of its crewmen, although modifications over the years had produced an improved vehicle that still had a respectable service life left. Since 1962, when production stopped, the airframe had received several strengthening modifications. More appropriate instrumentation had been installed for low-level missions although such rides were bumpy and uncomfortable for the crews (only high-altitude strikes were programmed for the skies over Iraq). The aircraft could carry cruise missiles under its wings, and its ability to accurately deliver its ordnance had been enhanced. It appeared to some that the Methuselah of the Strategic Air Command would be around forever: No comfort to the six-man crew that sat in cramped spaces for as long as 24 hours at a stretch.

The B-1 was conspicuous by its absence, its engines displaying temperament that caused an unacceptable rate of failure for long-range combat duty. The bat-shaped B-2 was still in the development and test stages, although its little brother, the spooky-looking F-117A Stealth fighter, was proving itself in the Iraqi theater of operations.

The naval component of USCENTCOM continued enforcing the sea embargo with some off-shore shelling and intensive carrier-launched air strikes. On Tuesday, February 19, the U.S.S. *Tripoli*, a helicopter assault ship, and the guided-missile cruiser U.S.S. *Princeton*, struck Iraqi mines in the Persian Gulf and suffered severe damage although neither ship was put out of operation. Both had to retire for repairs. The naval force continued to present a major amphibious and air-assault threat to Saddam's forces in Kuwait City.

Back in Washington the previous Sunday, February 17, President Bush had stated that Iraq's occupation of Kuwait would end "very, very soon." There was some concern over the visit of Saddam's foreign minister, Tariq Aziz, to Moscow. The earlier withdrawal offer by

Saddam had withered and died on the vine; but it could be that another, more palatable offering was in the making. The Soviets were not consulting with the other coalition partners on any political or diplomatic contacts with Iraq, but all were suspicious that Gorbachev was still trying for a unilateral peace coup.

He was, and the allies would soon learn of the new proposal from Baghdad.

CHAPTER 11

WITH THE RELATIVELY new AirLand Battle Doctrine as a guide, General Schwarzkopf's forces spent the final days before the attack going over tactical doctrine and sharpening up their communications and air-ground coordination. Some of the Forward Air Controllers (FACs) had switched to the lightning-fast F-16s from the slower, greater-risk A-10 Warthogs. Their response time would be faster and they could cover more ground in a shorter time. Even seconds could make a difference in a fast-developing combat situation. Once on the offensive, coalition forces would be moving hard and fast. The heavy Abrams tanks could hit forty miles per hour over decent terrain, and the desert provided a lot of that. The secret to the success of such a fast-moving campaign was the ability of the logistic forces to keep up with combat arms, and thousands of supply vehicles were loaded and ready. After all, an armored division could use more than a half-million gallons of fuel per day. If the fighting was intense, ammunition replenishment would be a priority item. It could be disastrous if the attacking forces outran their ammo, and the assault could easily stall.

The ability of logistic forces to keep up with the combat arms depended in turn upon the allies having complete air supremacy, which fortunately they already had. There would not be a single Iraqi aircraft to challenge the dash across the sands. That was the optimum condition, for in the open desert, supply vehicles would be visible for miles, their dust for scores of miles. However, dense desert dust thrown up by moving vehicles could also aid the attackers by augmenting the battle tactic of making smoke to hide troop movements.

The AirLand doctrine called for improved battlefield reconnaissance capabilities and more accurate means of determining location, whether yours, your other troops', or the enemies'. Consequently, E-8A JSTARS radar aircraft would be supplying realtime ground reconnaissance to commanders down to the battalion level. There would be two E-8As over the gulf, enabling the division and battalion commanders to "see" the picture (terrain, enemy disposition, effect of artillery fire) well in advance of their movement while the corps commander could concentrate his attention to more distant threats.

Troops, even down to the level of individuals in many cases, would have available small pocket receivers that indicated their exact position to within 75 feet by receiving data from navigation satellites dedicated to the Global Positioning System, a capability that would be particularly valuable in the featureless desert environment.

As the fifth week of DESERT STORM drew to a close, some troops in the field had a scare as a pair of Scuds broke up in the skies over the town of Hafar al Batin near the center of Schwarzkopf's long battle line paralleling the Saudi Arabia-Iraq-Kuwait border. A third missile continued and was intercepted by Patriots over Riyadh. Falling debris caused light injuries.

B-52s were dropping fuel-air bombs onto Iraqi minefields. The weapons released a number of smaller bomblets as they approached the ground which would explode and send out a wide pattern of vaporized fuel. A second explosion would ignite the fuel and create an explosion that would not only provide the crushing pressure needed to detonate mines, but would also suck up air to suffocate any troops that happened to be in the area. The ever-versatile C-130 Hercules also took a bombing role by delivering 15,000-pound "daisy cutter" bombs. These had been used in Vietnam to clear out large areas of forest, providing instant helicopter landing pads. In Iraq, the tremendous explosions were used to explode and destroy sections of the barrier minefields.

Figures varied and it was impossible to get an accurate count, but Iraqis were beginning to cross over the border with their hands held high and white pieces of cloth gripped in their fingers. Estimates by *USA Today* put the number at over 1,000, about ten percent of them officers. They were all fatigued and hungry and many slogged through the sands on cloth-wrapped feet. The men were fed and sent to the rear to join the others who had trickled in previously as enemy prisoners of war. The coalition forces felt little animosity toward the defecting Iraqis, knowing they were also just soldiers doing their job, and very poorly supported. The prevalent emotion of U.S. troops was pity. Most of the Iraqis, once they realized they were safe, smiled and enthusiastically nodded their gratitude. A few even embraced their captors.

With a few minor exceptions, the coalition forces were in position and ready. To the extreme east, along the coast, were two Saudi Arabian divisions and the 5th Marine Expeditionary Brigade. To their left was the 2d Marine Division. Continuing on west was the 1st Marine Division and the Tiger Brigade of the Army's 2d Armored Division. The three forces would be heading across Kuwait toward Kuwait City and its gulf port facilities.

Next (still working west) were the Egyptian-led Arab armored forces, and to their left were the U.S. 1st Cavalry Division and a brigade of the 2d Armored Division. They would charge directly across the border to reinforce the deception that coalition forces were conducting a standard head-on attack against Iraqi forces.

On the western side of the diamond-shaped neutral zone that straddled the Saudi-Iraqi border, were the British 1st Armored Division and the U.S. Army VII Corps (2d Armored Cavalry Regiment, 3d Armored Division, 1st Armored Division, and 1st Infantry Division). VII Corps would attack straight north into Iraq and once well inside, the British would veer right and confront the leading edge of the Republican Guards, while the remainder of VII Corps would speed across to execute a flanking action and attack the Guards from the southwest.

To VII Corps' left was the U.S. XVIII Airborne Corps with the 3d Armored Cavalry Regiment, the 24th Mechanized Infantry Division, and the 101st Airborne Division. The XVIII Airborne Corps would execute the "Hail Mary" plunge across Iraq to the Euphrates River, seal off the Republican Guards from the west, and cut off any Iraqi use of Highway 8.

Finally, to the extreme west, the 6th French Light Armored Division and the 2d Brigade of the U.S. 82d Airborne Division would strike across Iraq and provide protection against any Iraqi troops moving down from northern Iraq.

General Schwarzkopf had a few other aces up his sleeve. Two days before the ground war would begin, elements of the 1st Marine Division crossed into Kuwait and began barrier clearing operations. Approximately 3,000 marines moved ten miles into the interior and were mapping minefields and ordering artillery and air strikes onto Iraqi positions. They took prisoners and in general prepared the way for the 1st and 2d Marine Divisions. There were other "discreet" marine incursions, all designed to soften or eliminate barriers to the main thrust. Such advance tactics were politically sensitive and not revealed until after the cessation of hostilities.

Special Forces inside Iraq were extremely active, providing target information and monitoring Iraqi movements and communications.

Diplomatic efforts continued but they were primarily Iraqi and Soviet attempts to pull Saddam's chestnuts out of the fire with as little damage as possible. Saddam's original withdrawal plan had met with such overwhelming rejection that Foreign Minister Aziz had traveled to Moscow on February 17 (U.S. date) for talks with Gorbachev to discuss modifications of Saddam's original offer to withdraw. Two days later, Aziz returned to Baghdad with a modified plan that was little better.

It did provide for an Iraqi "unconditional withdrawal," but then stated conditions! Iraqi conditions, no less. Moscow would encourage a Middle East peace conference to consider the Palestinian question. Moscow would also: endeavor to protect Saddam from any war crimes prosecution; argue for the integrity of all Iraqi borders; and work for getting all economic sanctions lifted.

The immediate reaction of Americans who had been following diplomatic developments was, "Who appointed Gorbachev to handle everyone's affairs?"

The proposal also omitted any mention of three key matters. There was no indication of a withdrawal timetable. The date for returning POWs was not addressed. There were no statements relative to the restoration of the exiled government of Kuwait. And most objectionable, the proposal would allow Saddam Hussein to save face (most U.S. officials were determined that he should not only lose face but certain other parts of his figurative anatomy).

Bush, acting for the coalition, not only rejected the plan, but sent to Gorbachev allied conditions for any withdrawal plan, assuming correctly that Aziz would be running back to Moscow and Gorbachev could relay the coalition demands.

Aziz did hurry back, but the result was only another Soviet-sponsored proposal to which Iraq agreed. It still called for a prior cease-fire plus:

—Iraqi forces would have four days to withdraw from Kuwait City, three weeks to withdraw from Kuwait;
—POWs would be released three days after cease-fire;
—after withdrawal, all U.N. resolutions would be negated; and
—supervision of cease-fire would be by U.N.-selected, non-coalition forces

Bush received the plan in Washington on Thursday evening, February 21 (U.S. date). Earlier that day, Gorbachev had phoned and stated that he had given Aziz a new Soviet-proposed withdrawal plan, and that it was being transmitted to Washington. Bush immediately realized it was unsatisfactory, and handed the task of drafting a reply to a group of his top aides; the president continued on to the Ford Theater to watch the movie *Black Eagles*, a story of black airmen during World War II.

Bush's staff agreed that the Soviet proposal was obviously not acceptable. The coalition would set the conditions for withdrawal, and nothing else would be considered. Withdrawal from Kuwait City had to be completed within 48 hours; from all of Kuwait within one week. Withdrawal would be along allied-marked routes; any units straying from those routes or not withdrawing would be attacked. There would be an immediate exchange of prisoners and the Kuwaiti government would be immediately reinstalled.

All other provisions of the coalition's response to the first Soviet-sponsored plan remained, including the removal of all land mines and an immediate cessation of destruction within Kuwait.

Meanwhile, after serious discussion with Secretary Cheney, General Powell, and top White House aides, Bush had taken Powell's recommendation (concurred with by Cheney) and selected the deadline date for Saddam's withdrawal. Despite the flurry of Soviet activity, the coalition would wait no longer for Saddam to play his silly games. At noon EST on Saturday, February 23, if Iraq had not started a massive withdrawal from Kuwait, the opportunities were over. Coalition partners had all been contacted and agreed.

The final communiqué was sent to Gorbachev to relay to Aziz and Bush made the public announcement.

The ultimatum apparently had little effect upon Saddam. He condemned it as a "shameful ultimatum," and his forces in Kuwait punctuated his response with more oil facility fires and atrocities within Kuwait City. Saddam repeated his boast that Iraq was ready to fight any ground war.

It is difficult to imagine what Saddam's state of mind was at that moment. Did he have any awareness that, in the ground skirmishes to date, his forces had been soundly defeated? Did he know that already more that a thousand of his troops had surrendered? He had access to outside news broadcasts, although reports from the front were probably slow in reaching him despite the fact he still had radio communications. Were his commanders telling him everything? If so, did he realize the gravity of the situation? Did he already figure that defeat by the allies would result in their withdrawal in the end, but that he would be safe from any purging and remain the man who had taken on not only the Great Satan but a host of other powers? Whatever his reasoning, he foolishly stood his ground, perhaps trapped by his own ego.

It is not difficult to imagine General Schwarzkopf's state of mind. His forces had completed a masterful air campaign. He enjoyed the confidence of his superiors and his troops. He had given many hours to the construction of his ground battle plan. His field commanders were ready. His logistic supply lines were complete, protected, and ready to move with the fighting forces.

High noon on Saturday had a particularly fateful ring to it back in Washington, but in Schwarzkopf's headquarters it would be 8:00 P.M. on Saturday evening. Schwarzkopf had already directed authorization of advance movement for some tanks to enable them to break through the twelve-foot-high sand berms and clear a way for the main bodies of troops.

There would be little chance for Schwarzkopf and his staff to rest that night, for eight hours later coalition forces would move out on the largest land offensive since World War II.

CHAPTER 12

FOUR O'CLOCK IN the morning is an excellent time to attack. The enemy, particularly if he is anticipating an assault, has stayed awake most of the night, trying to be alert. By 4:00 AM, the body clock is calling for a rest period and there is a definite need for sleep. Reactions are sluggish. Senses are less alert. Small noises, for example, may pass unnoticed. If the trooper is dug in or in the confined quarters of a tank or other armored vehicle, his joints ache from lack of movement and he has that "cooped up" feeling, despite the fact that it has been his battle station for some time and he has trained under the same conditions. Why don't they do it? his mind asks. He wants to urinate more often than normal.

If the trooper is malnourished or poorly clothed for

the weather conditions, he is physically miserable, and the above conditions are aggravated.

The attacker may have similar problems, but they are probably not as severe. Knowing when he would be moving out, he has possibly been able to get some rest. Even fitful sleep is better than no sleep at all. As the moment approaches, his body will give him a shot of adrenaline. He will aggressively institute the action, not sleepily respond to it. He has the advantage from the physiological standpoint.

There is another aspect of an early morning attack and it goes all the way back to our childhood when things went bump in the night. To the defender, there is "something out there in the dark." He knows that it is something to fear. He wants to see what it is. To the attacker, the darkness is cover for his preliminary movement. This gives him the advantage from the psychological standpoint as well.

Finally, darkness can be used to add another frightening dimension to warfare. There is a terrible contrast between the darkness and the bright flashes of artillery fire and bomb or missile explosions that create an unearthly panorama of death and destruction. The sounds of combat are exaggerated. All of us have noticed that when we see a police car in the dead of night speeding through the dark streets with its red lights flashing and its siren wailing, we have feelings of dread. During daylight, the image is less fearful, as one subconsciously reasons that the "black and white" may very well be rushing to the scene of a minor accident; at night one is more prone to think that a crime has been committed, perhaps a violent robbery or a gruesome murder.

Iraqi positions had been pounded day and night by coalition gun and rocket attacks. Now, as it comes in all wars, the moment of the mudsoldier had arrived.

Seven thousand miles away, in the Pentagon War Room, no one was sleepy. True, it was only 8:00 PM, but it was a time of day for the body under normal conditions to relax. Initially, there had not been universal agreement among all of the top military staffers as to the date for beginning the land war. The Air Force would have liked to have seen a longer air phase, knowing that if Saddam capitulated before the AirLand phase began, air power would have achieved a great victory and set a new course for future warfare. The Army and Marines, however, were anxious to go with their traditional belief that the war wouldn't be over until enemy land and troops were *taken*, not just pounded to a pulp. Air power alone hadn't beaten the Japanese or the Germans. The North Koreans had not been held back and the North Vietnamese had not been deterred by any air forces (all admittedly ill-managed). As for Saddam, after 35 days of unopposed air war, he was still arrogant and his troops were still in place, although his war-making capability had been significantly reduced.

In the field, Lieutenant General Gus Pagonis (who had received a promotion during DESERT STORM, largely for his brilliant logistic performance) wanted a few more days to set his logistic bases in optimum position. Schwarzkopf was satisfied that sufficient support was already in place, an opinion that many in the administration shared. To go on with an air war, now

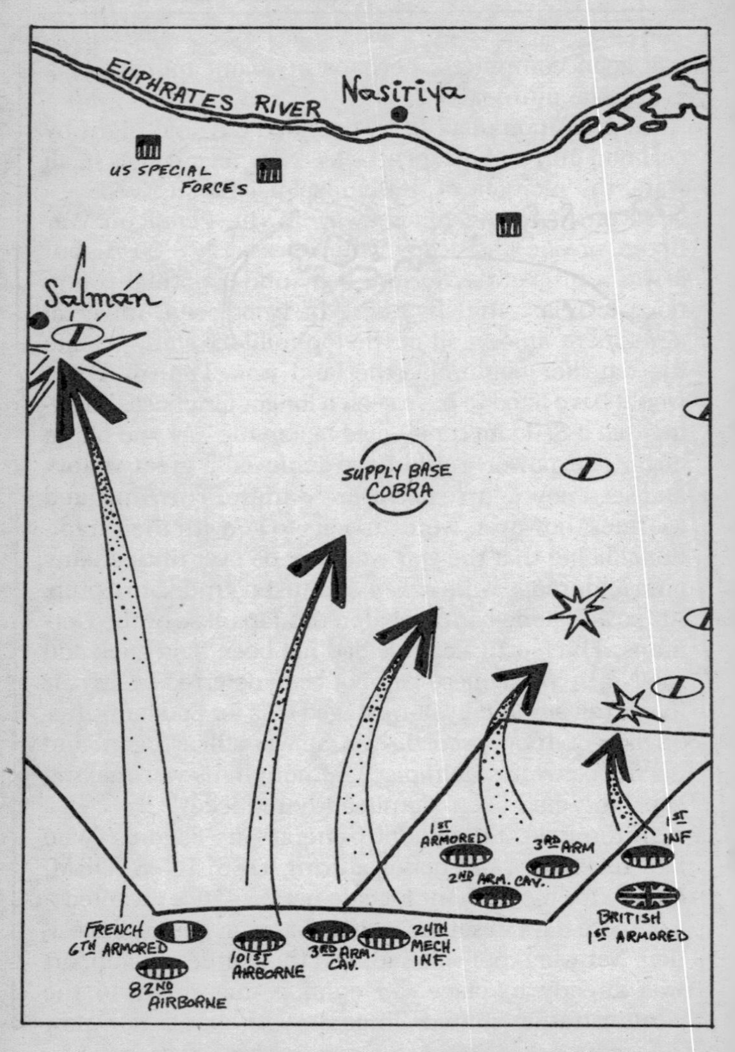

COALITION ATTACK ON KUWAIT AND IRAQ—
Sunday, February 24th 1991

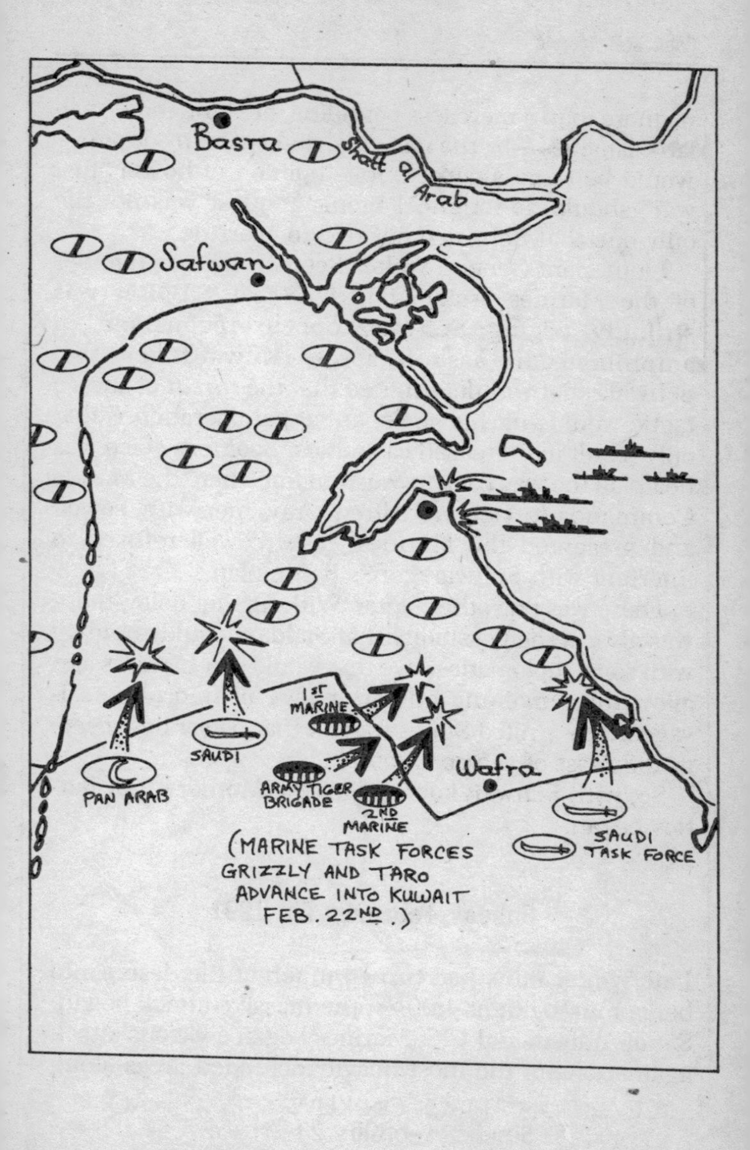
Basra
Shatt al Arab
Safwan
1ST MARINE
SAUDI
PAN ARAB
ARMY TIGER BRIGADE
2ND MARINE
Wafra
SAUDI TASK FORCE
(MARINE TASK FORCES GRIZZLY AND TARO ADVANCE INTO KUWAIT FEB. 22ND.)

no more than a merciless pounding, creating more civilian casualties—by the nature of things, not by design—would be to go against public opinion of how a "just war" should be fought. Pagonis' request was not the only one that Schwarzkopf had to overrule.

Lieutenant General Walter Boomer, the commander of the Marines assigned to DESERT STORM was strongly urging Schwarzkopf to reconsider an amphibious/air assault across Kuwaiti beaches. Schwarzkopf was determined that the *threat* of such a tactic would suit his goals; an actual operation would only result in increased casualties. Boomer's voice was heard all the way back in Washington where the Marine Commandant, General Alfred Gray, met with Powell and presented the Marines' case. Powell refused to interfere with Schwarzkopf's battle plan.

There was one other factor. With further delay, there was always the possibility that Saddam could come up with some diplomatic move that would suit his aims and allow the remaining military power of Iraq to be salvaged. The United States wanted that power destroyed, as did most of the coalition.

So, with Schwarzkopf's "Go!" the Mother of All Battles began.

Sunday, February 24, 1991

Late winter rains had turned much of the desert into beige mush, but at 4:00 AM the massive attack began. Saudi Arabian and U.S. Marines began a vicious attack against one of the most heavily defended areas along the Kuwaiti border, the coastal area that was the easternmost point of the attack.

The 1st and 2d Marine Divisions, led by Boomer himself, were nearby to the west and charged ahead in a brilliantly executed textbook operation that Schwarzkopf would later praise in the strongest of terms. Boomer's 17,000 offshore marines were still in their amphibious assault feint position, and naval gunfire was reinforcing the ruse that the marines were about to land. There was some serious concern about the trenches that the Iraqis had filled with oil, but coalition air had bombed many of these with napalm and burned off the oil. The Saudis did encounter flame barriers, but simply pushed sand into the trenches with their tanks and bulldozers and snuffed out the fires before riding across. In a matter of hours, the allies had completely breached the "formidable" barrier, and even bypassed some Iraqi units that were poised to fight but failed to mount a counterattack. Those that did were met and overpowered with concentrated artillery fire, tank attacks, and air strikes.

On the opposite end of the allied line, French Legionnaires and the U.S. 82d Airborne Division rapidly penetrated into Iraq, their first objective the Al Faman airfield and military installation at As Salman, 120 miles inside the border. They soon encountered Iraqi resistance, but U.S. artillery and French-built Gazelle helicopters with their HOT antitank missiles engaged and overcame the Iraqis. A large number of men surrendered.

Immediately to the east, after a short hold for bad weather, troop-carrying Chinook helicopters raced

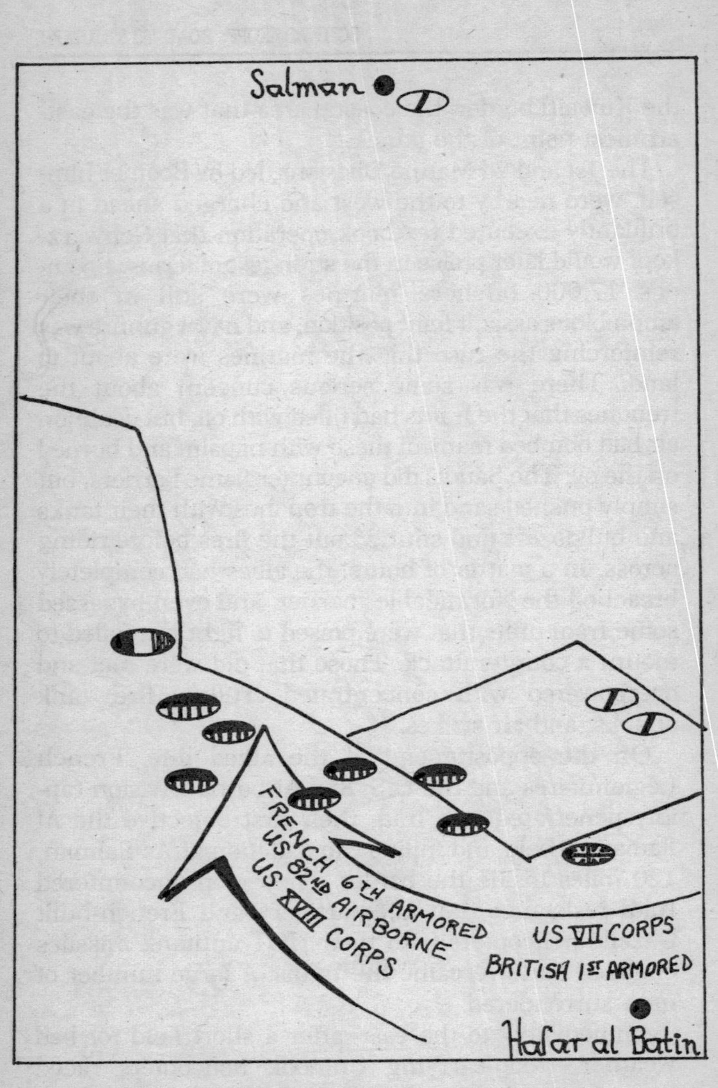

THE "HAIL MARY" STRATEGY—Sunday, February 24th 1991

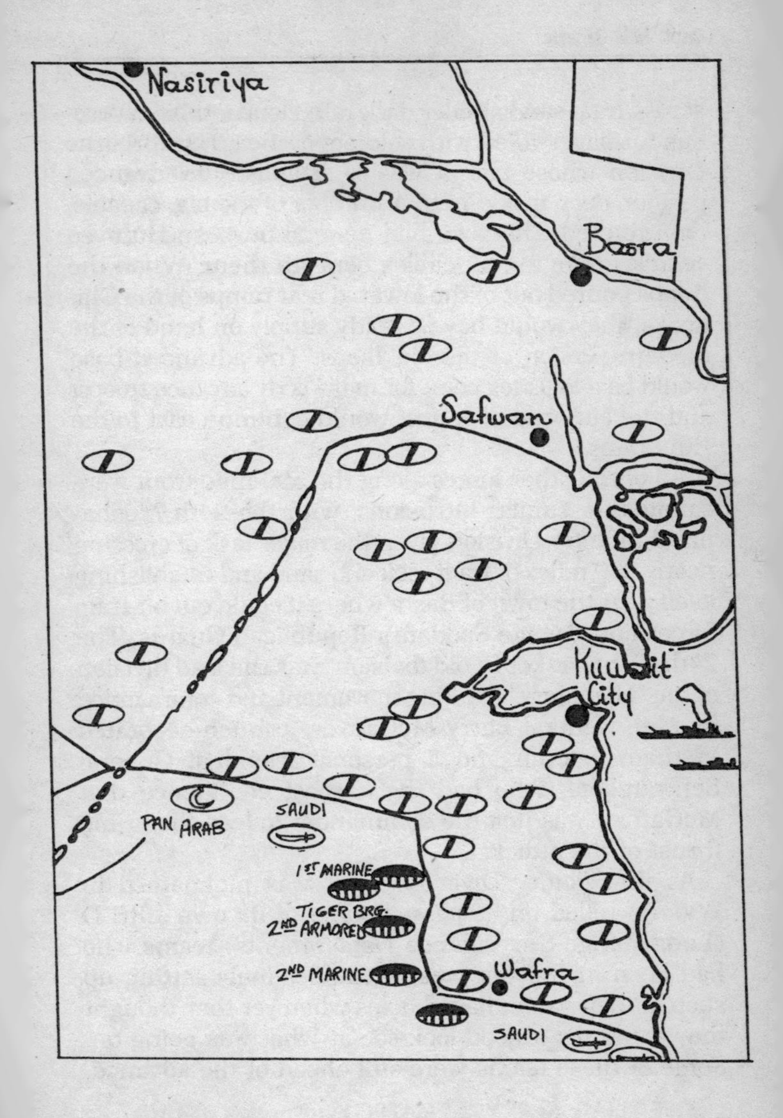

Nasiriya
Basra
Safwan
Kuwait City
PAN ARAB
SAUDI
1ST MARINE
TIGER BRG.
2ND ARMORED
2ND MARINE
Wafra
SAUDI

across Iraqi sands under dark rain clouds, their cavernous fuselages filled with soldiers of the 101st Airborne Division whose task it was to establish an advanced position deep inside Iraq. A number of the big, capable, twin-rotored Chinooks had general-purpose Humvee vehicles hanging on cables beneath them. When the troops poured out of the lowered rear ramps of the Chinooks, they would have a ready supply on hand of the modern version of the old Jeeps. The advanced base would be a logistics point for main-body advance troops, and the airborne task unit would continue east to the Euphrates.

All of the other forces along the 500-mile front were conducting similar intrusions, with the 24th Mechanized Infantry Division given the major task of crossing nearly 250 miles of Iraqi rock and sand and establishing itself near the town of Basra where it could cut off Iraqi forces and engage Saddam's Republican Guards. The 24th, Schwarzkopf's old division, was the lead division of the "Hail Mary" flanking movement and commanded by Maj. General Barry McGaffrey, a much-decorated Vietnam veteran and a personal friend of General Schwarzkopf, who had the highest confidence that McGaffrey was just the commander to lead the prime thrust of the attack.

As the Victory Division (as it was nicknamed in WW II) rolled on, it passed some of its own LRS-D (Long Range Surveillance Detachments) teams who had been in Iraq for weeks, clandestinely setting up shop in their camouflaged holes wherever they thought they could get a good look-see at what was going on. Some of these teams were still ahead of the advance, several hundred miles into the interior. Blackhawk Special Forces helicopters had inserted the six-man teams and left them to fend for themselves, a task at which they were very good. The LRS-D troopers dug down into the rocky soil and erected a covering over their hole made of chicken wire and desert cloth, and then covered that with soil and sand and rocks until it was practically invisible. One could walk right up to the position and not see it—and on several occasions wandering bedouin tribesmen did just that. With their HF "burst" transmissions, the Special Forces soldiers could send back valuable information in seconds that might take several hours to get through the spy satellites' communications networks.

Now, as the men and armor of the 24th went charging past, the "spies in the sand" were content to remain unnoticed. They would be recovered and sent to other locations as needed.

All coalition forces were advancing with unanticipated speed even though they had not expected major resistance early in the operation. As dawn came on the 24th, the advancing troops in the easternmost sector could see the eerie glow of oil fires in Kuwait. The scene was surreal, with hundreds of orange fires and rising columns of black smoke fanning out across the dawn sky, blown south by the prevailing winds. Back at DESERT STORM headquarters, Schwarzkopf and his staff were receiving early combat reports with great satisfaction. The operation had kicked off on time and indications were that almost all forces were ahead of schedule.

Secretary Cheney had issued a news blackout once the attack began and the briefings at Riyadh were temporarily suspended. Media pool members were allowed to accompany some troops, however, and would file detailed reports.

At 10:00 PM, EST, two hours after the first official troops crossed into Iraq and Kuwait, President Bush made a televised announcement that coalition troops had started their attack and that "the liberation of Kuwait had entered its final phase."

The 1st Cavalry Division, along with its brigade of 2d Armored Division troops, began its thrust up the Wadi al Batin corridor along the Kuwait-Iraq border. To the division's right, Arab forces charged in parallel to convince the Iraqis that the main attack was head-on toward their dug-in positions.

As the coalition forces advanced, more and more Iraqis surrendered and were herded to the rear as quickly as possible.

Disturbing reports were received back at General Schwarzkopf's headquarters that Iraqi troops in Kuwait City were beginning a new round of terrible atrocities, partly in retaliation against Free Kuwaitis who had used the invasion to increase their own resistance actions. More than ever, Schwarzkopf wanted to prevent the Iraqis from escaping to the north. Once Kuwait was liberated, there should be a reckoning for all of the crimes committed against Kuwait City and its inhabitants.

Very seldom do operations go as planned. There are almost always unforeseen contingencies that arise requiring a change in tactics. But the first day of the Air-Land war was going exactly as programmed. If anything, it was going better than expected. By evening, the 1st Marines had reached al Jaber airport, halfway to Kuwait City.

There is a great satisfaction in advancing into enemy territory and realizing that the softening-up process by artillery or rockets or aircraft or naval gunfire has done its job well. Your life and well-being is the payoff and although the DESERT STORM troops knew that the first hours were only the start, they began to take comfort in the number of Iraqis surrendering and the light armed resistance. It was almost too good to be true and for those precious few who had been casualties and those who would die and be wounded during the next 76 hours, it proved to be a false comfort.

Monday, February 25, 1991

All coalition forces pressed forward. The fast-moving French, way out on the left flank, had to be ordered to pause. General Schwarzkopf wanted to make sure that they did not outstrip the XVIII Airborne Corps to the east, as the French were their flank protection against reinforcing Iraqi troops coming down from the north. The French halted and dug in.

As the day wore on, the number of surrendering Iraqis began to present an unusual problem. There were so many of them that advancing forces were hard-pressed to release sufficient troopers to corral the Iraqis and herd them toward containment areas. The best estimates indicated that there were already some 20,000. The Iraqis had lost the will to fight. Many of their officers deserted early in the operation, even though special execution squads had been assigned by Saddam to shoot deserters. The Iraqis were tired and scared and in no mood to stand up to the awesome power that the coalition was throwing across Iraq and into Kuwait.

Eastward all down the battle line, it was the same story. The few Iraqis who did choose to fight were quickly vanquished. In some of the early tank skirmishes, the American M1A1s with their superior-range guns stayed just beyond the range of the shells thrown by the Soviet-made T-72s and T-62s and fired away at will until the enemy tanks were silenced.

From his Riyadh headquarters, General Schwarzkopf declared that the first day of battle had been a "dramatic success." He was careful to state that he did not regard the great mass of surrenders as signs that the Iraqis were poor fighters, just that they knew they had lost their cause.

Although Baghdad Radio was reporting that there were orders for Iraqi forces to withdraw, Saddam continued to urge his troops on, and was declaring over Baghdad Radio that Iraqi forces were inflicting severe casualties on the allied forces. He did have enough force left to continue his Scud launches, although they were very few—only two on Monday, both aimed at Israel, neither of which did any damage. It appeared that he was at the bottom of his stockpile since the Scuds he was using were breaking apart in midair under the stress of flight. Homemade, they were probably poorly constructed with weak welds and mismatched parts. The warheads could still be damaging, of course, as Schwarzkopf would tragically learn in the near future.

The British 1st Armored Division was shattering the minimal Iraqi resistance it encountered and running into the same dilemma of "too many" prisoners of war as the other coalition forces. Reports of the Brits' success had reached England and for the first time during her 39-year reign, Queen Elizabeth II made a wartime address, praising coalition forces for their success and reporting to her subjects that she was with them in their prayers for a speedy victory.

As the central forces continued their spectacular dash across southern Iraq, to the east the 1st Marines encountered serious but localized resistance in the Burgan oil field near Kuwait International Airport. With a massive, sudden, single volley of fire, the marines drove the Iraqis from their holes, where they were attacked by U.S.M.C. tanks and Cobra attack helicopters. The Iraqis were soundly defeated in the late afternoon battle, losing 50 of their 60 tanks to no losses for the marines. It was a day of complete success for the coalition forces.

The VII Corps was well into Iraq and along with troops of the XVIII Airborne Corps was preparing to take on the main body of Republican Guards. The British 1st Armored Division was also positioning itself to strike the Guards from the south. The "elite" Guards would not only have to combat the deadly U.S. Abrams but the low-slung British Challenger battle tank, a destroyer that military experts placed among the best in the world. Tough tank crewmen were known to have test-fired a quick six rounds at a target 1,100 meters away and with pinpoint accuracy scored hits so closely spaced that all could have gone through an open area the size of a large window. The dual thrust was a pincer that would leave the Guards no place to go but northeast toward Basra—and the Tigris River, or southeast, back into Kuwait, which was rapidly filling with the 1st Cavalry Division, a brigade of the 2nd Armored Division, and the Egyptian-led Arab forces.

And then something fearful happened. Out of the darkness, a Scud dropped from the cloudy sky. It broke up but its warhead struck an American barracks at the base near Dhahran in Saudi Arabia. Twenty-eight soldiers, including three women, died, and 90 were injured, some seriously. All of the deaths were reservists and the tragedy brought home the cruel fact that although front-line troops were advancing at an amazing rate, those in rear-area support were also at risk. Fate picked the barracks, not the Scud, but even a deliberate aim could not have caused more damage. General Schwarzkopf grieved not only for the young men and women who had gone to sleep that night feeling relatively secure, but for the families who had sent their children and spouses off to war. He had a special place in his huge heart for his men, but he also knew the terrible toll the loss of a soldier took on those back home. It was a hard time.

Tuesday, February 26, 1991

The Iraqis in Kuwait City had gotten the message and decided that discretion was the better part of valor. They began to flee the city, a number in their own military vehicles, others in whatever transportation they could commandeer or steal. It was a welcome exodus, for both Kuwaiti insurgents and allied troops had anticipated a bloody house-to-house mopping-up operation in Kuwait

City. The Iraqis left, but with a vengeance, destroying property at will and murdering still more citizens, while taking along a large number as hostages. The road north began to fill with vehicles of every description and appeared to be settling into "the mother of all traffic jams." Instead, it became a killing field as coalition aircraft swarmed down and fired at everything, whether it was moving or not. The biggest danger to the pilots was the possibility of running into each other. Vehicles on the ground exploded and caught fire. Lead tanks stalled, and crews evacuated and headed north on foot. It was a complete rout.

The Iraqis were desperately trying to reach the Euphrates River, despite the fact that the allies had been very thorough in taking out the bridges. Some escaping troops would manage to construct pontoon and boat bridges.

Brigadier General Neal had resumed daily briefings in Riyadh and stated that forces withdrawing with no arms or armor would not be attacked, but that all others were considered to be in a combat-withdrawal situation and would be fired on.

In the waters off Kuwait City, SEAL teams were active, removing mines and obstacles, keeping up the ruse that an amphibious landing was imminent. Naval ships maneuvered within sight of the shore troops. The Iraqis launched aged Chinese Silkworm antiship missiles, but the SSMs were shot down by fleet guns.

Baghdad Radio reported again that Saddam Hussein had ordered his troops to withdraw from Kuwait, and there were a few scattered voices in the world community urging a cease-fire. White House spokesman Mar-

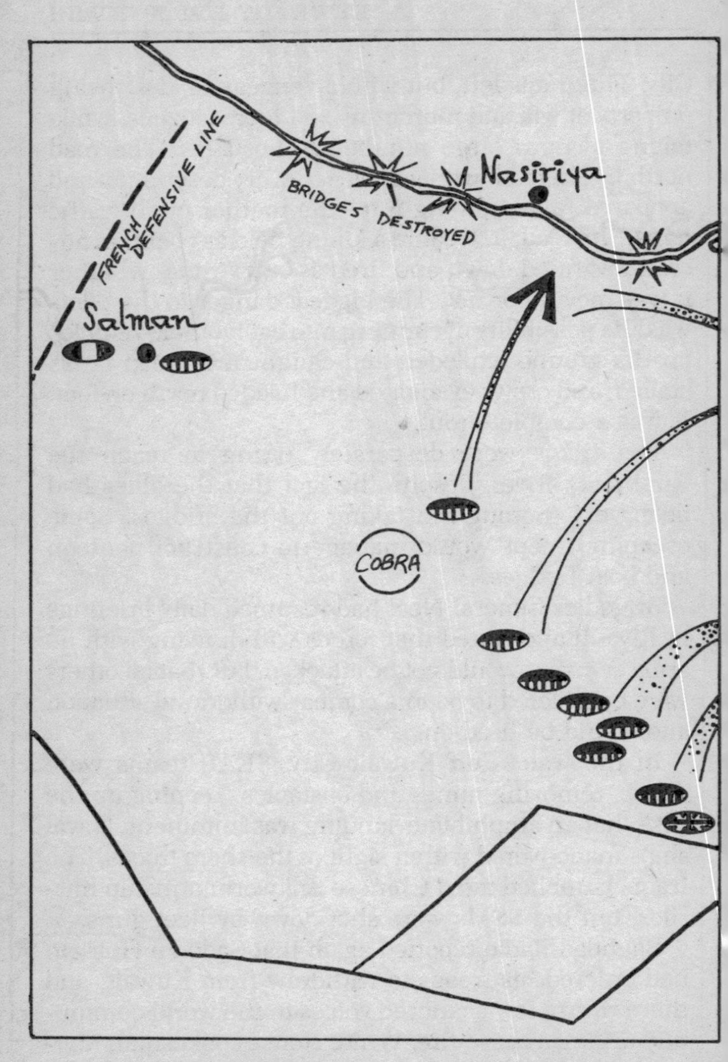

COALITION ATTACK ON KUWAIT AND IRAQ—
February 25th and 26th 1991

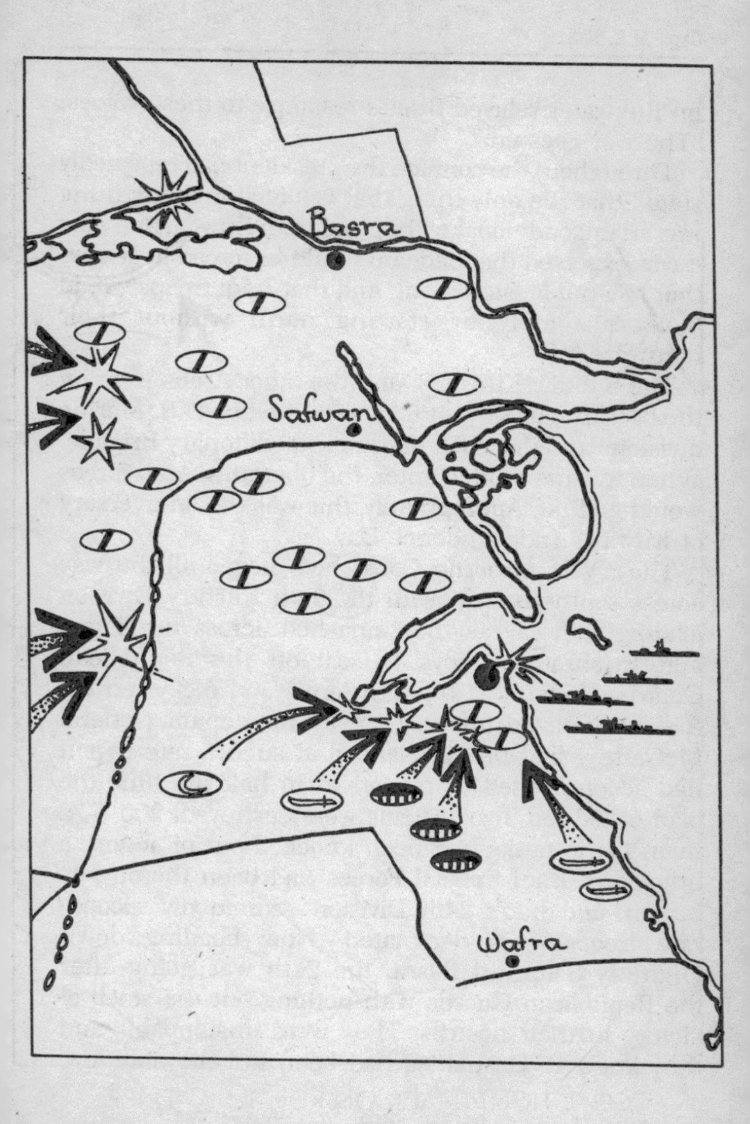
Basra
Safwan
Wafra

lin Fitzwater relayed Bush's response to these voices: "The war goes on."

Throughout the conflict, the president had repeatedly stated that the only thing that would stop the fighting was an unconditional withdrawal from Kuwait, and now it was expected that Saddam would be forced to declare that in a public statement, and that Iraqi troops would have to comply by starting north without their hardware.

The Saudis, Kuwaitis, and the other Arabs were on the outskirts of Kuwait City with the two U.S. Marine divisions and the Fifth Marine Expeditionary Brigade. When it came time to enter, the liberating Arab forces would go first. Appropriately, this was the anniversary of Kuwait's Independence Day.

The XVIII Airborne Corps had moved all the way across southern Iraq, with the 24th Infantry Division leading the way as they smashed across the Tigris and Euphrates valleys to seal off the Republican Guards. Undaunted by sandstorms and rocky terrain, the mechanized infantry led by Lieutenant General McGaffrey had plunged ahead at such a rate that it had accomplished its objectives in half the time the plan called for. Two airfields were destroyed, 200 T-72 main battle tanks had been knocked out of action, a brigade of Iraqi Special Forces had been thoroughly beaten, and Iraq's 24th Division—admittedly second-rate troops—was decimated. Now heading down Highway 8 toward Basra, the 24th was going after the Republican Guards with nothing but the smell of victory in their nostrils. They were unstoppable and they knew it. Casualties had been unbelievably low, with just six killed and 16 wounded. The Iraqis were no match for the 24th's 290 M1A1s, 270 Bradley fighting vehicles, 75 155-mm mobile howitzers, nine MRLSs (Multiple Rocket Launching Systems), and deadly Apache attack helicopters.

General McGaffrey had personally led his forces since the first hour and now as he rode with them south along Highway 8, they were encountering the unsuspecting fleeing Iraqis. It was impossible not to feel some pity for the ragged troops of Iraq, who had been placed in a position where there had been no hope.

Behind the 24th, a huge logistic force stretched all the way back along what had been designated Supply Route Yankee, past two divisional supply depots in central Iraq, to near Ash Shu'bah in northern Saudi Arabia. Along the way, stranded Iraqi fuel resources had been gladly added to the millions of gallons brought in by the supply tankers.

McGaffrey had conditioned his troops long before the ground war began by requiring them to wear their chemical warfare suits. They were no longer required although still kept at the ready, and the troopers who had not had an opportunity to bathe in weeks were a motley-looking crew, strangely deodorized by the charcoal filtering feature of the suits.

The 24th had probably the toughest assignment in the operation from the standpoint of distance and time to reach its objective and number of forces to move. But it also had the excellent leadership of a great general officer in McGaffrey, and a logistic supply line second to none ever deployed.

As the end of the third day of the assault approached,

it was apparent that the massive coalition forces had completely overwhelmed the undergunned, undersupplied, and undernourished Iraqis. But there was no overkill. Complete divisions of Iraqis could have been wiped out, but were instead given the opportunity to surrender.

Far to the east, the Tiger Brigade of the 2d Armored Division and the 1st Marines were fighting similar battles as they approached Kuwait City. The Tiger Brigade had in their heritage the leadership of George Patton, and their task was to curl north around Kuwait City and cut off any retreat. The burning oil wells were pouring smoke over the entire area, severely reducing visibility. The Iraqis were waiting for them with dug-in positions and half-buried T-55 tanks, but they were no match for the 120-mm guns of Tiger armor, and Iraqi turrets were blown clear of their tank bodies with relentless accuracy. There were secondary explosions and tracers stitching the area all around the advance. As everywhere, small white pieces of cloth appeared, followed by one or more surrendering Iraqis. The men of the Tiger Brigade were careful to cease fire when that happened, but when they approached a vehicle, particularly one of the T-55s, the Iraqi crew had only a very short time to make their intention to surrender known; otherwise, they died in their vehicle. The fighting was intense and the gory sights of battle—dismembered bodies, charred corpses, and burning ripped steel—would remain with the men of the Tiger Brigade long after they left for home. Beside them, the 1st Marines were encountering similar resistance but with equal success and as the day wound down, soldiers and marines stood shoulder to shoulder at the gates to Kuwait City.

Armchair generals would soon be coming up with statements like, "Schwarzkopf completely overestimated the enemy forces," and, "It wouldn't have gone so well had Nazis or Russians been in those enemy tanks." How sad that the pundits missed the point. The low casualties were exactly the goal General Schwarzkopf had set for his plan. True, they were even lower than he'd imagined possible, but that was not the result of "overestimation." It was the result of a set of circumstances that could come together only once in a very great while: a great plan, comprehensive intelligence, extensive softening-up, superior arms, professional forces, and superb execution. As for the caveat about "Russians in those enemy tanks," there *had* been Soviet-trained crews inside, and Soviet tactics had been used. Admittedly, the Soviets might have used them better, but that was beside the point. Even so, the Soviet generals would later direct a new look at their tank fighting doctrine.

Saddam was desperately casting about for any chance of avoiding further humiliation. Foreign Minister Aziz hastily announced that Iraq would withdraw from Kuwait in accordance with the U. N. resolution; later he stated that several other resolutions would be complied with—but not all. Stubborn to the end. His statements did not even stimulate a coalition reply.

Wednesday, February 27, 1991

The noose was tightening around both Kuwait and Kuwait City. British units were astraddle the main Kuwait City–Basra highway. Beyond them, coalition forces had a multilayered blockade of troops and armor all the way from the southwestern border of Iraq to the Tigris.

The day was marked, however, by two large tank encounters with fierce fighting on both sides. The first took place at the Kuwait International Airport where marines ran up against a major Iraqi armored force. The smoke-filled sky stank and blocked out the sun with such efficiency flashlights had to be used to read maps and charts. The Iraqis fought with a tenacity seldom seen in the previous days of the operation, but the marines were by far the superior fighters and by the conclusion of the battle all 100 Iraqi tanks had been destroyed. One factor was the superior low-light sighting ability of the U.S. armor. Another was their superior maneuverability and firing range.

The second battle, much larger, took place along the Iraq-Kuwait border between American-British forces and the Iraqis. The engagement had been joined as early as Monday when the allies went up against the Republican Guard, and developed over the next 48 hours as the classic tank battle of the campaign. But the Yanks and Brits had a decided advantage with their M1A1s and Challengers. The full fury of the battle lasted all day Wednesday, and despite the length of the engagement and the ferocity with which it was fought in the final hours, it was really no contest. The Iraqis were fighting without air cover, they had very poor communications, and they were under continual air attack. In the final stages, the allies were chasing the Iraqis and destroying them almost at will. Later on, when General Schwarzkopf gave his briefing of the ground battle, wherein he referred to some 3,000 of Iraq's 4,700 tanks as being destroyed, he mentioned that another 700 should be added to the total as the result of Wednesday's two battles.

Some would have said, "It's all over but the shouting," but there was still a considerable amount of mopping-up to do and, as the military would say, "policing the battlefield." Certainly, consolidating control of Kuwait City was the first consideration for the remainder of the operation, and marines had already followed Arab forces in. The U.S. Embassy was being secured. There was some concern that the Iraqis had booby-trapped it or even that there could be Iraqi holdouts inside.

The Kuwaitis were ecstatic as they regained control of their capital city, and there were crowds of shouting and happy people in the streets. All were expressing their admiration and gratitude to the coalition forces, some of the resistance fighters firing their weapons into the air. They established their headquarters in the city, and for the next days they would be filtering out Iraqi collaborators and sympathizers for trial and punishment.

As Brigadier General Neal had announced earlier in the day, the Iraqis were in full retreat. Saddam announced that his troops were in complete withdrawal. There was a difference between the two state-

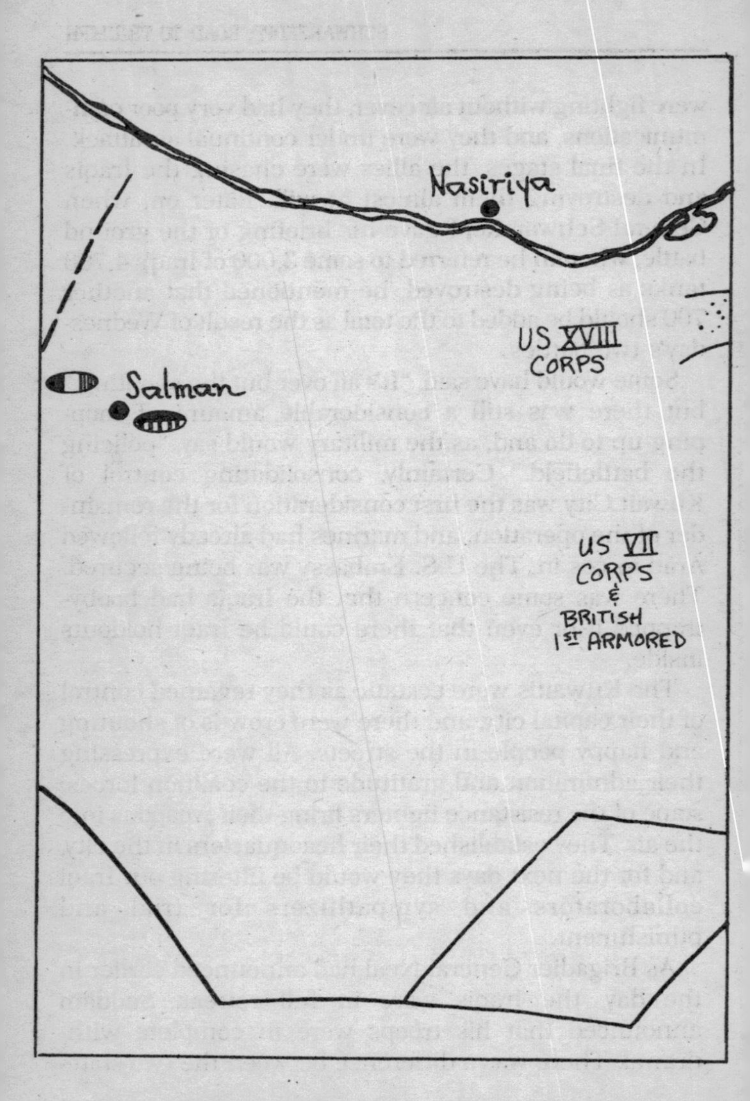

FINAL COALITION ATTACK—February 27th 1991

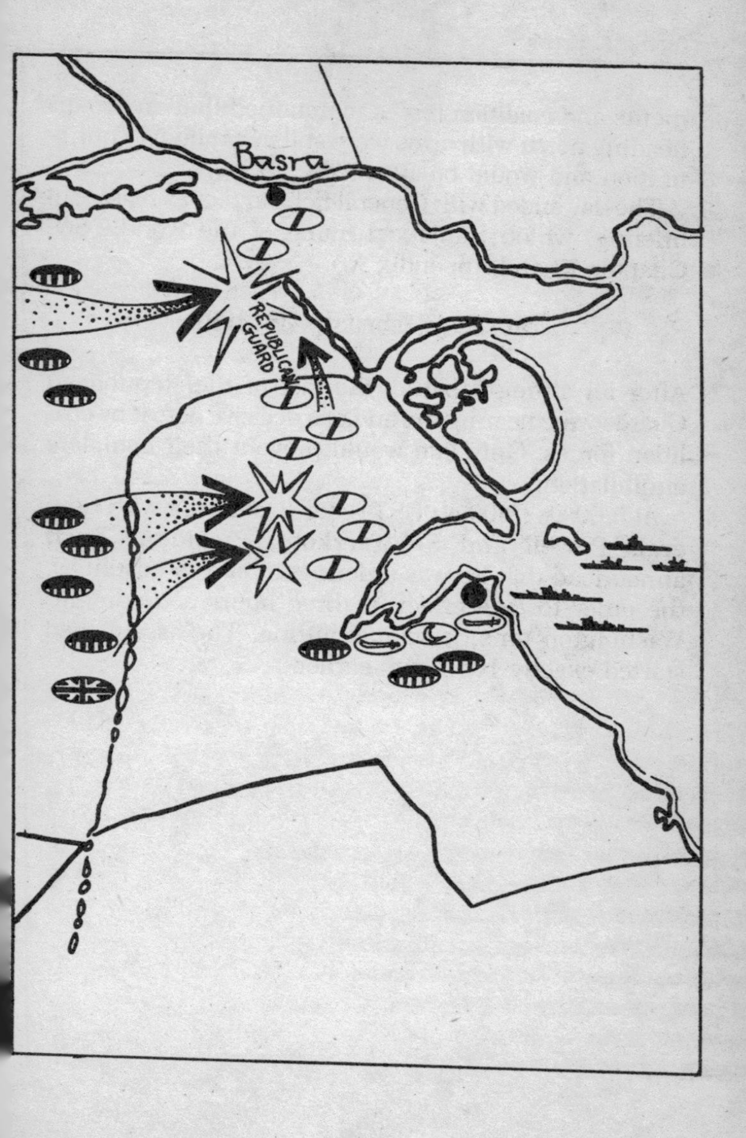
Basra
REPUBLICAN GUARD

ments and coalition forces maintained that any Iraqis heading north with arms were still in a military configuration and would be attacked.

The day ended with General Schwarzkopf's masterful briefing, which transfixed much of the world. (See Chapter 13 and Appendix A.)

Thursday, February 28, 1991

After an all-night battle, the last of the Republican Guards were nearing the end of a decisive defeat by coalition forces. Only fate would prevent their complete annihilation.

At 5:00 AM (9:00 PM EST), after conferring with Generals Powell and Schwarzkopf, President Bush announced that he was ordering an end to hostilities, the order to take effect in three hours at midnight, Washington time. It seemed fitting. The assault had started exactly 100 hours earlier.

CHAPTER 13

GENERAL H. NORMAN SCHWARZKOPF, United States Army, Commander in Chief, United States Central Command and Commander, Allied Forces, Operation DESERT SHIELD/STORM, walked out before the seated media correspondents at the general's headquarters in Riyadh, Saudi Arabia, a collapsible pointer in his right hand. He took his place behind a waist-high podium and faced his audience. American and Saudi Arabian flags were behind him, one on each side. To his right was an easel, upon which rested a small stack of rigid charts, and an officer assistant stood off to the left.

The Bear was about to tell it like it happened. It was 9:00 PM, Wednesday, February 27, 1991.

The general was in combat fatigues, the same uniform he and his troopers had worn since August 7, 1990. His were clean and precisely pressed and his trouser cuffs were tucked neatly into beige desert boots. On each side of his pointed collar were four black stars arranged in vertical rows. The black-embroidered Combat Infantryman Badge and Master Jump Wings were over his left shirt pocket, as was a sewn-on white strip carrying black stenciled letters: U. S. ARMY. A similar strip was over his right pocket. It said simply: SCHWARZKOPF. A watch was on each wrist, one on local time, the other on Eastern Standard Time.

The large body inside the uniform was about to burst its seams with satisfaction and pride, but the desert-camouflaged shirt and trousers were tailored to preserve the military look as General Schwarzkopf remained casually standing, facing his expectant audience. His presence dominated the briefing room; his personality held everyone's attention.

"Good evening, ladies and gentlemen," he began. After a brief introductory remark, he had his assistant uncover the first chart, and for the next 21 minutes, he detailed the 100-Hour War.

There was no bragging, no bravado, no gloating, no I-told-you-so attitude. Instead, it was a standard military briefing that used the series of charts to expand and clarify his statements and emphasize the tactics and progress of the operation.

It was his time of triumph, the climax of a series of briefings his command had provided for the past six and a half months. Follow-up briefings would be continued on a smaller scale until the business of USCENTCOM in Saudi Arabia was finished.

Referring to the chart, he talked of how, at the beginning of the deployment, he had brought in and positioned his forces in a defensive posture, and how that posture had been strengthened in November as a response to a marked increase in Iraqi forces facing coalition troops across the Saudi Arabian border.

With sincerity, he paid tribute to the naval forces that were present in the Persian Gulf and the Red Sea and spoke of their importance to the integrity of his command.

He made the observation that as the concept of his mission changed from defensive to offensive, he began to tally up the numbers and compare them to textbook treatments of troop strength when formulating an attack.

As he spoke, he moved slowly back and forth between the podium and the easel and used the extended pointer to trace the positions and movements of troops. Coalition forces were designated by blue rectangles; Iraqis by green, yellow, and red, depending upon how much attrition they had suffered. Maneuvers were indicated by large arrows.

He emphasized his three deceptive tactics: the amphibious threat, the hidden movement of large forces to the west for the wide flanking attack, and the massing of forces directly across from Iraqi strength in Kuwait. Considering the number of troops and the distance involved, several hundred miles, the undetected swing of forces to the west may even have been historic. In any event, it was a classic military ploy and would not have been possible except for the tremendous logistic support he had at his command. When the AirLand campaign started, General Schwarzkopf had 60 days of logistic supplies at forward depots.

The big threat was the Republican Guard, "hiding" over to the northeast near Basra in southern Iraq, and Schwarzkopf's plan was designed to hit them simultaneously from several directions with awesome force.

As he spoke, he did so with authority, for he had lived every moment of the campaign, keeping the scores of details straight in his mind and thinking ahead for alternative courses of action should the unexpected occur.

It hadn't.

The operation was proceeding as planned. As General Schwarzkopf continued, he singled out his various forces for special praise. The Marines and Saudis on the coastal attack performed magnificently and breached the formidable physical barriers with such success that Schwarzkopf was at a loss for words to describe the brilliant attack that saw the marines smashing into Kuwait through minefields and razor wire, and the Saudis fearlessly crossing flaming oil barriers. Military students would be studying the thrust for decades to come.

The other Arabs, under Egyptian command, had plunged straight into the heart of Iraqi resistance and on to Kuwait City.

VII Corps, XVIII Airborne Corps, and the 1st Cavalry Divison (with its brigade from 2d Armored Division) had exceeded all expectations. The British troops under VII Corps had performed superbly in their attack on the Republican Guards and the French had raced across Iraq side by side with troops of the 82d Airborne Division, both forces at great risk if Iraqi troops had poured down from the north.

As was his style, General Schwarzkopf made certain that all members of the international coalition and the three U.S. military services were acknowledged for their contribution to a precedent-setting operation. He made a point of singling out the daring Special Forces teams who had been placed deep in Iraqi territory prior to the ground attack, though he preferred not to go into any detail.

He reported with obvious satisfaction that casualties had been unbelievably low, but stressed that "the loss of even one life is intolerable to those of us in the military." At the time of his briefing, there were 79 killed in action, 213 wounded, and 44 missing in action (The final total of KIAs would be 129). At that moment, whether he intended to or not, he revealed to those present the very reason why a senior military commander is so carefully selected. A commander's most precious assets in battle are his men and women, and he must make the most difficult of decisions, that of sending his young people into harm's way. To General Schwarzkopf, every casualty was a personal tragedy, and that aspect of his personality endeared him not only to his troops but to the families of those troops. Schwarzkopf had risked his life for his troopers in Vietnam and when asked about it during a television interview subsequent to the Persian Gulf war, he would reply only that he "had no choice." They had been his men and his responsibility. In his mind his act had not been one of bravery. He had been scared stiff. Such is the stuff of which military leaders are made.

At the conclusion of his briefing, he accepted questions and fielded them with the same expertise he had shown in making his presentation. His answer to one question, "What are your impressions of Saddam Hussein as a military strategist?" brought forth first a Schwarzkopf "Ha!", then laughter, and finally a studied answer. As he spoke, he used the fingers of his left hand to count off Saddam's "attributes."

"As far as Saddam Hussein being a great military strategist, he is neither a strategist, nor is he schooled in the operational arts, nor is he a tactician, nor is he a general, nor is he a soldier. Other than that, he is a great military man. I want you to know that."

It became one of the most quoted of Schwarzkopf's remarks.

When asked further about special operations, he gave a guarded answer, but did admit that special forces were used to call in air strikes and perform strategic reconnaissance. A Special Forces team accompanied each Arab force at battalion level to assist in language problems and call in air support. Special forces also conducted the difficult and hazardous combat search-and-rescue missions. Enough said, he implied.

One questioner aroused old and terrible memories when he asked if maybe coalition forces had overrated Iraqi defensive barriers. The extensive Iraqi minefields immediately leaped to Schwarzkopf's mind and the general was back in Vietnam in the middle of his own minefield with his horribly wounded men.

Instantly, he became The Bear—The Grizzly Bear. He leaned forward across the podium, his jaw set, and his eyes burned down on the hapless correspondent.

"Have you ever been in a minefield?" he asked, perhaps wondering how *anyone* could think that a barrier featuring land mines, razor wire, trenches of burning oil, and covered by prepositioned artillery fires could *ever* be overrated.

"No," the reporter replied after an embarrassed pause.

"All there's got to be is *one* mine, and that's intense. . . ." Schwarzkopf had not meant to belittle the man and went on to answer the full question with an explanation of the hazards of the Iraqi barrier.

When questioned about the friendly fire that killed nine British soldiers, Schwarzkopf was unable to conceal the personal agony the terrible mistake had caused. It had been the second such incident, following the one that cost 11 U. S. marines their lives at the battle for Khafji just before the ground war began. Schwarzkopf knew he could not explain such a thing. In close combat, there was always the danger of losses to friendly fire. All any commander could do was work to prevent it.

The old question of body counts resurfaced.

"I don't think there's ever been . . . even in the history of warfare, been a successful count of the dead . . . The people who will know (the count) best, unfortunately, are the families that won't see their loved ones come home."

Fifty-one minutes after he had walked out, General Schwarzkopf thanked his audience and strode back to his war room. His forces were still fighting the Republican Guards near Basra. There was still much to do.

General Schwarzkopf's briefing, even though the 100-Hour War was still eleven hours away from the cease-fire, covered the entire DESERT SHIELD/

STORM operation in a complete but general manner. It would be weeks before operational reports were filed and military analysts could begin studying the details of minute-to-minute operations.

Several things were readily apparent, however. There was a miraculously low loss of life—later, some observers would refer to the loss-kill ratio as one of "biblical proportions," as if a divine presence had interceded on behalf of the "good guys." In any case, DESERT STORM had to be one of the most effective campaigns ever mounted.

The offensive capability of Iraqi military forces was being destroyed. Barring rearmament, Iraq would not present a military threat for years to come, even if Saddam Hussein remained in power.

The United States had purged its Vietnam complex. It *could* participate in a victorious war; it had led this one. And coalition partners, particularly the British, French and Saudis, demonstrated that their soldiers were as tough as any in the world.

There was a new relationship between the West and a number of the Arab nations. Hopefully, with a common goal, they could work together toward other things. Whether that ability to cooperate foreshadowed further achievements relative to the serious problems remaining in the Middle East remained to be seen. But the prospects were better. It appears that the coalition may have won the "peace after the war" that was one of the hopes of General Schwarzkopf.

One final tribute must be made to the coalition forces that joined the United States in defeating the Iraqi forces. The war against Saddam Hussein was an *allied* effort and there will be those in the future who will describe it as an American war. That is not true. America was just fortunate to have a large part of the resources and the trust of old and new allies, and the coalition put aside small differences for the common good of the free world. True, historians will include in their account of these times the gratitude the United States has earned for this worldwide vote of confidence and support. A new spirit has come out of the Persian Gulf conflict, and a renewed realization that the responsibility for peace on our planet is shared by the staunchest of friends.

As for General Schwarzkopf himself, he has emerged as a new American, even international, hero. But that should not be too surprising. The country desperately needed one and it needed a military figure to erase fifteen years of bad dreams. The Persian Gulf war, with its clear-cut mandate by the peoples of the world, was a timely vehicle to enable General Schwarzkopf to serve his country one last time before considering retirement. It gave him the opportunity to demonstrate his military leadership genius, his diplomatic discretion, and his complete understanding of the political factors that underlie every military conflict.

A professional soldier—or sailor or marine or airman—knows that his ultimate worth is his performance in combat. It is the only guide against which to measure his love of country and devotion to duty. When he has spent a lifetime either fighting or preparing to fight—even though he hates battle as no one else can hate it—and he has placed the welfare of those in his care before his own life, he has served with great honor.

Duty, honor, country. A magnificent calling.

General Schwarzkopf will not ever refer to himself as a hero. As he puts it, the heroes were the thousands of men and women who were his troops. But the general has had his days in the dark valley and has known so well the conflicting emotions of bone-chilling fear and undeniable responsibility that are the marks of a hero.

Maybe in his eyes we do him a disservice by honoring him so. Perhaps we should be content to look upon him as a soldier who saw his duty and did it as best he could. That would place him in an honorable category with millions of other Americans. He would most probably like that.

But destiny has robbed him of that modest goal. He must live the rest of his life in the bright glow of his superb achievement. H. Norman Schwarzkopf is now a permanent part of history.

And we are grateful.

•

APPENDIX A

Transcript: Briefing by General Schwarzkopf at Riyadh on Wednesday, February 27, 1991.

CENTCOM News Briefing
General H. Norman Schwarzkopf, USA
Riyadh, Saudi Arabia
Wednesday, February 27, 1991—1:00 P.M. (EST)

General Schwarzkopf: Good evening, ladies and gentlemen. Thank you for being here.

I promised some of you a few days ago that as soon as the opportunity presented itself I would give you a complete rundown on what we were doing, and more importantly, why we were doing it—the strategy behind what we were doing. I've been asked by Secretary Cheney to do that this evening, so if you will bear with me, we're going to go through a briefing. I apologize to the folks over here who won't be able to see the charts, but we're going to go through a complete briefing of the operation.

This goes back to 7 August through 17 January. As you recall, we started our deployment on the 7th August. Basically what we started out against was a couple of hundred thousand Iraqis that were in the Kuwait theater of operations. I don't have to remind you all that we brought over, initially, defensive forces in the form of the 101st, the 82d, the 24th Mechanized Infantry division, the 3d Armored Cavalry, and in essence, we had them arrayed to the south, behind the Saudi task force. Also there were Arab forces over here in this area, arrayed in defensive positions. That, in essence, is the way we started.

In the middle of November, the decision was made to increase the force because, by that time, huge numbers of Iraqi forces had flowed into the area, and generally in the disposition as they're shown right here. Therefore, we increased the forces and built up more forces.

I would tell you that at this time we made a very deliberate decision to align all of those forces within the boundary looking north towards Kuwait—this being King Khalid Military City over here. So we aligned those forces so it very much looked like they were all aligned directly on the Iraqi position.

We also, at the time, had a very active naval presence out in the gulf, and we made sure that everybody understood about that naval presence. One of the reasons why we did that is it became very apparent to us early on that the Iraqis were quite concerned about an amphibious operation across the shores to liberate Kuwait—this being Kuwait City. They put a very, very heavy barrier of infantry along here, and they proceeded to build an extensive barrier that went all the way across the border, down and around and up the side of Kuwait.

Basically, the problem we were faced with was this: When you looked at the troop numbers, they really outnumbered us about three-to-two, and when you consider the number of combat service support people we have—that's logisticians and that sort of thing in our armed forces, as far as fighting troops, we were really outnumbered two-to-one. In addition to that, they had 4,700 tanks versus our 3,500 when the buildup was complete, and they had a great deal more artillery than we do.

I think any student of military strategy would tell you that in order to attack a position you should have a ratio of approximately three-to-one in favor of the attacker.

In order to attack a position that is heavily dug in and barricaded such as the one we had here, you should have a ratio of five-to-one in the way of troops in favor of the attacker. So you can see basically what our problem was at that time. We were outnumbered as a minimum, three-to-two, as far as troops were concerned; we were outnumbered as far as tanks were concerned, and we had to come up with some way to make up the difference.

I apologize for the busy nature of this chart, but I think it's very important for you to understand exactly what our strategy was. What you see here is a color coding where green is a go sign or a good sign as far as our forces are concerned; yellow would be a caution sign; and red would be a stop sign. Green represents units that have been attritted below 50 percent strength; the yellow are units that are between 50 and 75 percent strength; and of course the red are units that are over 75 percent strength.

What we did, of course, was start an extensive air campaign, and I briefed you in quite some detail on that in the past. One of the purposes, I told you at that time, of that extensive air campaign was to isolate the Kuwaiti theater of operations by taking out all of the bridges and supply lines that ran between the north and the southern part of Iraq. That was to prevent reinforcement and supply coming into the southern part of Iraq and the Kuwaiti theater of operations. We also conducted a very heavy bombing campaign, and many people questioned why the extensive bombing campaign. This is the reason why. It was necessary to reduce these forces down to a strength that made them weaker, particularly along the front line barrier that we had to go through.

We continued our heavy operations out in the sea because we wanted the Iraqis to continue to believe that we were going to conduct a massive amphibious operation in this area. I think many of you recall the number of amphibious rehearsals we had, to include Imminent Thunder, that was written about quite extensively for many reasons. But we continued to have those operations because we wanted him to concentrate his forces—which he did.

I think this is probably one of the most important parts of the entire briefing I can talk about. As you know, very early on we took out the Iraqi Air Force. We knew that he had very, very limited reconnaissance means. Therefore, when we took out his air force, for all intents and purposes, we took out his ability to see what we were doing down here in Saudi Arabia. Once we had taken out his eyes, we did what could best be described as the "Hail Mary play" in football. I think you recall when the quarterback is desperate for a touchdown at the very end, what he does is he sets up behind the center, and all of a sudden, every single one of his receivers goes way out to one flank, and they all run down the field as fast as they possibly can and into the end zone, and he lobs the ball. In essence, that's what we did.

When we knew that he couldn't see us any more, we did a massive movement of troops all the way out to the west, to the extreme west, because at that time we knew that he was still fixed in this area with the vast majority of his forces, and once the air campaign started, he would be incapable of moving out to counter this move, even if he knew we made it. There were some additional troops out in this area, but they did not have the capability nor the time to put in the barrier that had been described by Saddam Hussein as an absolutely impenetrable tank barrier that no one would ever get through. I believe those were his words.

So this was absolutely an extraordinary move. I must tell you, I can't recall any time in the annals of military history when this number of forces have moved over this distance to put themselves in a position to be able to attack. But what's more important, and I think it's very, very important that I make this point, and that's these logistics bases. Not only did we move the troops out there, but we literally moved thousands and thousands of tons of fuel, of ammunition, of spare parts, of water, and of food out here in this area, because we wanted to have enough supplies on hand so if we launched this, if we got into a slug fest battle, which we very easily could have gotten into, we'd have enough supplies to last for 60 days. It was an absolutely gigantic accomplishment, and I can't give credit enough to the logisticians and the transporters who were able to pull this off, for the superb support we had from the Saudi government, the literally thousands and thousands of drivers of every national origin who helped us in this move out here. And of course, great credit goes to the commanders of these units who were also able to maneuver their forces out here and put them in this position.

But as a result, by the 23d of February, what you found is this situation. The front lines had been attritted down to a point where all of these units were at 50 percent or below. The second level, basically, that we had to face, and these were the real tough fighters we were worried about right here, were attritted to someplace between 50 and 75 percent. Although we still had the Republican Guard located here and here, and part of the Republican Guard in this area—they were very strong, and the Republican Guard up in this area strong, and we continued to hit the bridges all across this area to make absolutely sure that no more reinforcements came into the battle. This was the situation on the 23d of February.

I shouldn't forget these fellows. That SF stands for special forces. We put special forces deep into the enemy territory. They went out on strategic reconnaissance for us, and they let us know what was going on out there. They were the eyes that were out there, and it's very important that I not forget those folks.

This was the morning of the 24th. Our plan initially had been to start over here in this area, and do exactly what the Iraqis thought we were going to do, and that's take them on head-on into their most heavily defended area. Also, at the same time, we launched amphibious feints and naval gunfire in this area, so that they continued to think we were going to be attacking along this coast, and therefore, fixed air forces in this position. Our hope was that by fixing the forces in this position and with this attack through here in this position, we would basically keep the forces here, and they wouldn't know what was going on out in this area. I believe we succeeded in that very well.

At 4 o'clock in the morning, the Marines, the 1st

Marine Division and the 2d Marine Division, launched attacks through the barrier system. They were accompanied by the U.S. Army Tiger Brigade of the 2d Armored Division. At the same time, over here, two Saudi task forces also launched a penetration through this barrier. But while they were doing that, at four o'clock in the morning over here, the 6th French Armored Division, accompanied by a brigade of the 82d Airborne, also launched an overland attack to their objective up in this area, Al Faman Airfield, and we were held up a little bit by the weather, but by eight o'clock in the morning, the 101st Airborne air assault launched an air assault deep into enemy territory to establish a forward operating base in this location right here. Let me talk about each one of those moves.

First of all, the Saudis over here on the east coast did a terrific job. They went up against the very, very tough barrier systems; they breached the barrier very, very effectively; they moved out aggressively; and continued their attacks up the coast.

I can't say enough about the two Marine divisions. If I used words like brilliant, it would really be an underdescription of the absolutely superb job that they did in breaching the so-called impenetrable barrier. It was a classic, absolutely classic, military breaching of a very, very tough minefield, barbed wire, fire trenches type barrier. They went through the first barrier like it was water. They went across into the second barrier line, even though they were under artillery fire at the time—they continued to open up that breach. Then they brought both divisions streaming through that breach. Absolutely superb operation, a textbook, and I think it will be studied for many, many years to come as the way to do it.

I would also like to say that the French did an absolutely superb job of moving out rapidly to take their objective out here, and they were very, very successful, as was the 101st. Again, we still had the special forces located in this area.

What we found was, as soon as we breached these obstacles here and started bringing pressure, we started getting a large number of surrenders. I think I talked to some of you about that this evening when I briefed you on the evening of the 24th. We finally got a large number of surrenders. We also found that these forces right here, were getting a large number of surrenders and were meeting with a great deal of success.

We were worried about the weather. The weather was going to get pretty bad the next day, and we were worried about launching this air assault. We also started to have a huge number of atrocities of really the most unspeakable type committed in downtown Kuwait City, to include reports that the desalinization plant had been destroyed. When we heard that, we were quite concerned about what might be going on. Based upon that, and the situation as it was developing, we made the decision that rather than wait the following morning to launch the remainder of these forces, that we would go ahead and launch these forces that afternoon.

This was the situation you saw the afternoon of the 24th. The Marines continued to make great progress going through the breach in this area, and were moving rapidly north. The Saudi task force on the east coast was also moving rapidly to the north and making very, very good progress. We launched another Egyptian/Arab force in this location, and another Saudi force in this location—again, to penetrate the barrier. But once again, to make the enemy continue to think that we were doing exactly what he wanted us to do, and that's make a headlong assault into a very, very tough barrier system—a very, very tough mission for these folks here. But at the same time, what we did is continued to attack with the French; we launched an attack on the part of the entire 7th Corps where the 1st Infantry Division went through, breached an obstacle and minefield barrier here, established quite a large breach through which we passed the 1st British Armored Division. At the same time, we launched the 1st Armored Division, and the 3d Armored Division, and because of our deception plan and the way it worked, we didn't even have to worry about a barrier, we just went right around the enemy and were behind him in no time at all, and the 2d Armored Cavalry Division. The 24th Mech Division was also launched out here in the far west. I ought to talk about the 101st, because this is an important point.

Once the 101st had their forward operating base established here, they then went ahead and launched into the Tigris and Euphrates valleys. There are a lot of people who are still saying that the object of the United States of America was to capture Iraq and cause the downfall of the entire country of Iraq. Ladies and gentlemen, when we were here, we were 150 miles away from Baghdad, and there was nobody between us and Baghdad. If it had been our intention to take Iraq, if it had been our intention to destroy the country, if it had been our intention to overrun the country, we could have done it unopposed, for all intents and purposes, from this position at that time. That was not our intention, we have never said it was our intention. Our intention was truly to eject the Iraqis out of Kuwait and destroy the military power that had come in here.

So this was the situation at the end of February 24th in the afternoon.

The next two days went exactly like we thought they would go. The Saudis continued to make great progress up on the eastern flank, keeping the pressure off the Marines on the flank here. The special forces went out and started operating small boat operations out in this area to help clear mines, but also to threaten the flanks here, and to continue to make them think that we were, in fact, going to conduct amphibious operations. The Saudi and Arab forces that came in and took these two initial objectives turned to come in on the flank heading towards Kuwait City, located right in this area here. The British UK passed through and continued to attack up this flank. Of course, the VII Corps came in and attacked in this direction shown here. The 24th Infantry Division made an unbelievable move all the way across into the Tigris and Euphrates valley, and proceeded in blocking this avenue of egress out, which was the only avenue of egress left because we continued to make sure that the bridges stayed down. So there was no way out once the 24th was in this area, and the 101st continued to operate in here. The French, having succeeded in achieving all their objectives, then set up a flanking position, a flank guard position here, to make sure there were no forces that could come in and get us from the flank.

By this time we had destroyed, or rendered completely ineffective, over 21 Iraqi divisions.

Of course, that then brings us to today. Where we are today is we now have a solid wall across the north of the 18th Airborne Corps consisting of the units shown right here, attacking straight to the east. We have a solid wall here, again of the VII Corps also attacking straight to the east. The forces that they are fighting right now are the forces of the Republican Guard.

Again, today we had a very significant day. The Arab forces coming from both the west and the east closed in and moved into Kuwait City where they are now in the process of securing Kuwait City entirely, and ensuring that it's absolutely secure. The 1st Marine Division continues to hold Kuwait International Airport. The 2d Marine Division continues to be in a position where it blocks any egress out of the city of Kuwait, so no one can leave. To date, we have destroyed over 29—destroyed or rendered inoperable—I don't like to say destroyed because that gives you visions of absolutely killing everyone, and that's not what we're doing. But we have rendered completely ineffective over 29 Iraqi divisions. The gates are closed. There is no way out of here, there is no way out of here, and the enemy is fighting us in this location right here.

We continue, of course, high level air power. The air has done a terrific job from the start to finish in supporting the ground forces, and we also have had great support from the Navy—both in the form of naval gunfire and in support of carrier air.

That's the situation at the present time.

Peace is not without a cost. These have been the U.S. casualties to date. As you can see, these were the casualties we had in the air war; then of course, we had the terrible misfortune of the Scud attack the other night which, again, because the weapon malfunctioned, it caused death, unfortunately, rather than in a proper function. Then, of course, these are the casualties in the ground war, the total being as shown here.

I would just like to comment briefly about that casualty chart. The loss of one human life is intolerable to any of us who are in the military. But I would tell you that casualties of that order of magnitude considering the job that's been done and the number of forces that were involved is almost miraculous, as far as the light number of casualties. It will never be miraculous to the families of those people, but it is miraculous.

This is what's happened to date with the Iraqis. They started out with over 4,000 tanks. As of to date, we have over 3,000 confirmed destroyed—and I do mean destroyed or captured. As a matter of fact, that number is low because you can add 700 to that as a result of the battle that's going on right now with the Republican Guard. So that number is very, very high, and we've almost completely destroyed the offensive capability of the Iraqi forces in the Kuwaiti theater of operations. The armored vehicle count is also very very high, and of course, you can see we're doing great damage to the artillery. The battle is still going on, and I suspect that these numbers will mount rather considerably.

I wish I could give you a better number on this, to be very honest with you. This is just a wild guess. It's an estimate that was sent to us by the field today at noontime, but the prisoners out there are so heavy and so extensive, and obviously, we're not in the business of going around and counting noses at this time to determine precisely what the exact number is. But we're very, very confident that we have well over 50,000 prisoners of war at this time, and that number is mounting on a continuing basis.

I would remind you that the war is continuing to go on. Even as we speak right now there is fighting going on out there. Even as we speak right now there are incredible acts of bravery going on. This afternoon we had an F-16 pilot shot down. We had contact with him, he had a broken leg on the ground. Two helicopters from the 101st, they didn't have to do it, but they went in to try and pull that pilot out. One of them was shot down, and we're still in the process of working through that. But that's the kind of thing that's going on out on that battlefield right now. It is not a Nintendo game—it is a tough battlefield where people are risking their lives at all times. There are great heroes out there, and we ought to all be very, very proud of them.

That's the campaign to date. That's the strategy to date. I'd now be very happy to take any questions anyone might have.

Q: I want to go back to the air war. The chart you showed there with the attrition rates of the various forces was almost the exact reverse of what most of us thought was happening. It showed the front line troops attritted to 75 percent or more, and the Republican Guard, which a lot of public focus was on when we were covering the air war, attritted less than 75. Why is that? How did it come to pass?

A: Let me tell you how we did this. We started off, of

course, against the strategic targets. I briefed you on that before. At the same time, we were hitting the Republican Guard. But the Republican Guard, you must remember, is a mechanized armor force for the most part, that is very, very well dug in, and very, very well spread out. So the initial stages of the game, we were hitting the Republican Guard heavily, but we were hitting them with strategic-type bombers rather than pinpoint precision bombers.

For lack of a better word, what happened is the air campaign shifted from the strategic phase into the theater. We knew all along that this was the important area. The nightmare scenario for all of us would have been to go through, get hung up in this breach right here, and then have the enemy artillery rain chemical weapons down on troops that were in a gaggle in the breach right there. That was the nightmare scenario. So one of the things that we felt we must have established is an absolute, as much destruction as we could possibly get, of the artillery, the direct support artillery, that would be firing on that wire. That's why we shifted it in the very latter days, we absolutely punished this area very heavily because that was the first challenge. Once we got through this and were moving, then it's a different war. Then we're fighting our kind of war. Before we get through that, we're fighting their kind of war, and that's what we didn't want to have to do.

At the same time, we continued to attrit the Republican Guard, and that's why I would tell you that, again, the figures we're giving you are conservative, they always have been conservative. But we promised you at the outset we weren't going to give you anything inflated, we were going to give you the best we had.

Q: He seems to have about 500–600 tanks left out of more than 4,000, as just an example. I wonder if in an overview, despite these enormously illustrative pictures, you could say what's left of the Iraqi Army in terms of how long could it be before he could ever be a regional threat, or a threat to the region again?

A: There's not enough left at all for him to be a regional threat to the region, an offensive regional threat. As you know, he has a very large army, but most of the army that is left north of the Tigris/Euphrates valley is an infantry army, it's not an armored army, it's not an armored heavy army, which means it really isn't an offensive army. So it doesn't have enough left, unless someone chooses to re-arm them in the future.

Q: You said the Iraqis have got these divisions along the border which were seriously attritted. It figures to be about 200,000 troops, maybe, that were there. You've got 50,000 prisoners. Where are the rest of them?

A: There were a very, very large number of dead in these units—a very, very large number of dead. We even found them, when we went into the units ourselves, we found them in the trench lines. There were very heavy desertions. At one point we had reports of desertion rates of more than 30 percent of the units that were along the front here. As you know, we had quite a large number of POWs that came across, so I think it's a combination of desertions, of people that were killed, of the people that we've captured, and of some other people who are just flat still running.

Q: It seems you've done so much, that the job is effectively done. Can I ask you, what do you think really needs more to be done? His forces are, if not destroyed, certainly no longer capable of posing a threat to the region. They seem to want to go home. What more has to be done?

A: If I'm to accomplish the mission that I was given, and that's to make sure that the Republican Guard is rendered incapable of conducting the type of heinous acts that they've conducted so often in the past, what has to be done is these forces continue to attack across here, and put the Republican Guard out of business. We're not in the business of killing them. We have pay ops aircraft up. We're telling them over and over again, all you've got to do is get out of your tanks and move off, and you will not be killed. But they're continuing to fight, and as long as they continue to fight, we're going to continue to fight with them.

Q: That move on the extreme left which got within 150 miles of Baghdad, was it also a part of the plan that the Iraqis might have thought it was going to Baghdad, and would that have contributed to the deception?

A: I wouldn't have minded at all if they'd gotten a little bit nervous about it. I mean that, very sincerely. I would have been delighted if they had gotten very, very nervous about it. Frankly, I don't think they ever knew it was there. I think they never knew it was there until the door had already been closed on them.

Q: I'm wondering how much resistance there still is in Kuwait, and I'm wondering what you would say to people who would say the purpose of this war was to get the Iraqis out of Kuwait, and they're now out. What would you say to the public that is thinking that right now?

A: I would say there was a lot more purpose to this war than just get the Iraqis out of Kuwait. The purpose of this war was to enforce the resolutions of the United Nations. There are some twelve different resolutions of the United Nations, not all of which have been accepted by Iraq to date, as I understand it. But I've got to tell you, that in the business of the military, of a military commander, my job is not to go ahead and at some point say that's great, they've just now pulled out of Kuwait—even though they're still shooting at us, they're moving backward, and therefore, I've accomplished my mission. That's not the way you fight it, and that's not the way I would ever fight it.

Q: You talked about heavy press coverage of Imminent Thunder early on, and how it helped fool the Iraqis into thinking that it was a serious operation. I wondered if you could talk about other ways in which the press contributed to the campaign. (Laughter)

A: First of all, I don't want to characterize Imminent Thunder as being only a deception, because it wasn't. We had every intention of conducting amphibious operations if they were necessary, and that was a very, very real rehearsal—as were the other rehearsals. I guess the one thing I would say to the press that I was delighted with is in the very, very early stages of this operation when we were over here building up, and we didn't have very much on the ground, you all had given us credit for a whole lot more over here. As a result, that gave me quite a feeling of confidence that we might not be attacked quite as quickly as I thought we were going to be attacked. Other than that, I would not like to get into the remainder of your question.

Q: What kind of fight is going on with the Republican Guard? And is there any more fighting going on in Kuwait, or is Kuwait essentially out of the action?

A: No. The fight that's going on with the Republican Guard right now is just a classic tank battle. You've got fire and maneuver, they are continuing to fight and shoot at us as our forces move forward, and our forces are in the business of outflanking them, taking them to the rear, using our attack helicopter, using our advanced technology. I would tell you that one of the things that has prevailed, particularly in this battle out here, is our technology. We had great weather for the air war, but right now, and for the last three days, it's been raining out there, it's been dusty out there, there's black smoke and haze in the air. It's an infantryman's weather—God loves the infantryman, and that's just the kind of weather the infantryman likes to fight in. But I would also tell you that our sights have worked fantastically well in their ability to acquire, through that kind of dust and haze, the enemy targets. The enemy sights have not worked that well. As a matter of fact, we've had several anecdotal reports today of enemy who were saying to us that they couldn't see anything through their sights and all of a sudden, their tank exploded when their tank was hit by our sights. So that's one of the indications [we look for].

Q: Are you saying . . .
A: A very, very tough air environment, obviously, as this box gets smaller and smaller, and the bad weather, it gets tougher and tougher to use the air, and therefore, the air is acting more in an interdiction role than any other.

Q: Can you tell us why the French, who went very fast in the desert in the first day, stopped [inaudible] and were invited to stop fighting after 36 hours?
A: That's not exactly a correct statement. The French mission on the first day was to protect our left flank. What we were interested in was making sure we confined this battlefield—both on the right and the left—and we didn't want anyone coming in and attacking these forces, which was the main attack, coming in from their left flank. So the French mission was to go out and not only seize Al Salman, but to set up a screen across our left flank, which was absolutely vital to ensure that we weren't surprised. So they definitely did not stop fighting. They continued to perform their mission, and they performed it extraordinarily well.

Q: The Iraqi Air Force disappeared very early in the air war. There was speculation they might return and provide cover during the ground war. Were you suspecting that? Were you surprised they never showed themselves again?
A: I was not expecting it. We were not expecting it,

but I would tell you that we never discounted it, and we were totally prepared in the event it happened.

Q: Have they been completely destroyed? Where are they?

A: There's not an airplane flown. A lot of them are dispersed throughout civilian communities in Iraq. We have proof of that.

Q: How many divisions of the Republican Guard now are you fighting, and any idea how long that will take?

A: We're probably fighting on the order of . . . There were a total of five of them up here. One of them we have probably destroyed yesterday. We probably destroyed two more today. I would say that leaves us a couple that we're in the process of fighting right now.

Q: Did you think this would turn out, I realize a great deal of strategy and planning went into it, but when it took place, did you think this would turn out to be such an easy cake walk as it seems? And secondly, what are your impressions of Saddam Hussein as a military strategist? (Laughter)

A: First of all, if we thought it would have been such an easy fight, we definitely would not have stocked 60 days' worth of supplies on these log bases. As I've told you all for a very, very long time, it is very, very important for a military commander never to assume away the capabilities of his enemy. When you're facing an enemy that is over 500,000 strong, has the reputation they've had of fighting for eight years, being combat-hardened veterans, has a number of tanks and the type of equipment they had, you don't assume away anything. So we certainly did not expect it to go this way.

As far as Saddam Hussein being a great military strategist, he is neither a strategist, nor is he schooled in the operational arts, nor is he a tactician, nor is he a general, nor is he a soldier. Other than that, he's a great military man. I want you to know that. (Laughter)

Q: I wonder if you could tell us anything more about Iraqi casualties on the battlefield; you said there were large numbers. Are we talking thousands, tens of thousands? Any more scale you can give us?

A: I wish I could answer that question. You can imagine, this has been a very fast-moving battle, as is desert warfare. As a result, even today when I was asking for estimates, every commander out there said we just can't give you an estimate. It went too fast, we've gone by too quickly.

Q: Very quickly, the special operations folks—could you tell us what their front role was?

A: We don't like to talk a lot about what the special operations do, as you're well aware. But in this case, let me just cover some of the things they did. First of all, with every single Arab unit that went into battle, we had special forces troops with them. The job of those special forces was to travel and live right down at the battalion level with all those people to make sure they could act as the communicators with friendly English-speaking units that were on their flanks, and they could also call in air strikes as necessary, they could coordinate helicopter strikes, and that sort of thing. That's one of the principal roles they played, and it was a very, very important role. Secondly, they did a great job in strategic reconnaissance for us. Thirdly, the special forces were 100 percent in charge of the combat search and rescue, and that's a tough mission. When a pilot gets shot down out there in the middle of nowhere, surrounded by the enemy, and you're the folks that are required to go in and go after them, that is a very tough mission, and that was one of their missions. Finally, they also did some direct action missions, period.

Q: General, there have been reports that when the Iraqis left Kuwait City they took with them a number of the Kuwaiti people as hostages. What can you tell us about this?

A: We've heard that they took up to 40,000. I think you've probably heard the Kuwaitis themselves who were left in the city state that they were taking people, and that they have taken them. So I don't think there's any question about the fact that there was a very, very large number of young Kuwaiti males taken out of that city within the last week or two. But that pales to insignificance compared to the absolutely unspeakable atrocities that occurred in Kuwait in the last week. They're not a part of the same human race, the people that did that, that the rest of us are. I've got to pray that that's the case.

Q: Can you tell us more about that?

A: No sir, I wouldn't want to talk about it.

Q: Could you give us some indication of what's happening to the forces left in Kuwait? What kind of forces are they, and are they engaged at the moment?

A: You mean these up here?

Q: The ones in Kuwait, the three symbols to the right.

A: I'm not even sure they're here. I think they're probably gone. We picked up a lot of signals with people, there's a road that goes right out here and goes out that way, and I think they probably, more than likely, are gone. So what you're really faced with is you're ending up fighting the Republican Guard heavy mech and armor units that are there. Basically what we want to do is capture their equipment.

Q: They're all out of Kuwait then?

A: I can't say that. I wouldn't be the least bit surprised if there are not pockets of people all around here who are just waiting to surrender as soon as somebody uncovers them and comes to them, but we're certainly not getting any internal fighting going on across our lines of communication or any of that sort of thing.

Q: General, not to take anything away from the Army and the Marines on the breaching maneuvers . . .

A: I hope you don't.

Q: But many of the reports from the pools we've gotten from your field commanders and the soldiers were indicating that these fortifications were not as intense or as sophisticated as they were led to believe. Is this a result of the pounding that they took that you described earlier, or were they perhaps overrated in the first place?

A: Have you ever been in a minefield?

Q: No.

A: All there's got to be is one mine, and that's intense. There were plenty of mines out there, plenty of barbed wire. There were fire trenches, most of which we set off ahead of time. But there were still some that were out there, the Egyptian forces had to go through fire trenches. There were a lot of booby traps, a lot of barbed wire—not a fun place to be. I've got to tell you probably one of the toughest things that anyone ever has to do is to go up there and walk into something like that and go through it, and consider that while you're going through it and clearing it, at the same time you're probably under fire by enemy artillery. That's all I can say.

Q: Was it less severe than you had expected? You were expecting even worse, in other words.

A: It was less severe than we expected, but one of the things I contribute that to is the fact that we went to extensive measures to try and make it less severe. We really did. I didn't mean to be facetious with my answer, but I've just got to tell you that that was a very tough mission for any person to do, particularly in a minefield.

Q: Is the Republican Guard your only remaining military objective in Iraq? I gather there have been some heavy engagements. How would you rate this army you face—from the Republican Guard on down?

A: Rating an army is a tough thing to do. A great deal

of the capability of an army is its dedication to its cause and its will to fight. You can have the best equipment in the world, you can have the largest numbers in the world, but if you're not dedicated to your cause, if you don't have the will to fight, then you're not going to have a very good army.

One of the things we learned right prior to the initiation of the campaign, that of course contributed, as a matter of fact, to the timing of the ground campaign, is that so many people were deserting, and I think you've heard this, that the Iraqis brought down execution squads whose job was to shoot people in the front lines. I've got to tell you, a soldier doesn't fight very hard for a leader who is going to shoot him on his own whim. That's not what military leadership is all about. So I attribute a great deal of the failure of the Iraqi Army to fight, to their own leadership. They committed them to a cause that they did not believe in. They all are saying they didn't want to be there, they didn't want to fight their fellow Arabs, they were lied to, they were deceived when they went into Kuwait, they didn't believe in the cause, and then after they got there, to have a leadership that was so uncaring for them that they didn't properly feed them, they didn't properly give them water, and in the end, they kept them there only at the point of a gun.

The Republican Guard is entirely different. The Republican Guard are the ones that went into Kuwait in the first place. They get paid more, they got treated better, and oh by the way, they also were well to the rear so they could be the first ones to bug out when the battlefield started folding, while these poor fellows up here who didn't want to be here in the first place, bore the brunt of the attack. But it didn't happen.

Q: Can you tell us something about the British involvement, and perhaps comment on today's report of ten dead through friendly fire?

A: The British, I've got to tell you, have been absolutely superb members of this coalition from the outset. I have a great deal of admiration and respect for all the British that are out there, and particularly General Sir Peter [Delabiyea] who is not only a great general, but he's also become a very close personal friend of mine. They played a very, very key role in the movement of the main attack. I would tell you that what they had to do was go through this breach in one of the tougher areas, because I told you they had reinforced here, and there were a lot of forces here, and what the Brits had to do was go through the breach and then fill up the block, so the main attack could continue on without forces over here, the mechanized forces over here, attacking that main attack in the flank. That was a principal role of the British. They did it absolutely magnificently, and then they immediately followed up in the main attack, and they're still up there fighting right now. So they did a great job.

Q: The 40,000 Kuwaiti hostages taken by the Iraqis. Where are they right now? That's quite a few people. Are they in the line of fire? Do we know where they are?

A: No, no. We were told, and a lot of this is anecdotal. We were told that they were taken back to Basra. We were also told that some of them were taken all the way back to Baghdad. We were told 100 different reasons why they were taken. Number one, to be a bargaining chip if the time came when bargaining chips were needed. Another one was for retribution because, of course, at that time Iraq was saying that these people were not Kuwaitis, these were citizens of Iraq and therefore, they could do anything they wanted to with them. So I just pray that they'll all be returned safely before long.

Q: The other day on television, the Deputy Soviet Foreign Minister said, they were talking again already about re-arming the Iraqis. There's some indication that the United States, as well, needs to have a certain amount of armament to retain a balance of power. Do you feel that your troops are in jeopardy finishing this off, when already the politicians are talking about re-arming the Iraqis? How do you feel about that?

A: I certainly don't want to discuss [inaudible] because that's way out of my field. I would tell you that I'm one of the first people that said at the outset that it's not in the best interest of peace in this part of the world to destroy Iraq, and I think the President of the United States has made it very clear from the outset that our intention is not to destroy Iraq or the Iraqi people. I think everyone has every right to legitimately defend themselves. But the one thing that comes through loud and clear over, and over, and over again to the people that have flown over Iraq, to the pilots that have gone in against their military installations, when you look at the war machine that they faced, that war machine definitely was not a defensive war machine, and they demonstrated that more than adequately when they overran Kuwait and then called it a great military victory.

Q: Before starting the land phase, how much were you concerned by the Iraqi planes coming back from Iran? And do we know what happened to the Iraqi helicopters?

A: As I said before, we were very concerned about the return of the Iraqi planes from Iran, but we were prepared for it. We have been completely prepared for any type of air attack the Iraqis might throw against us, and oh, by the way, we're still prepared for it. We're not going to let down our guard for one instant, so long as we know that capability is there, until we're sure this whole thing is over.

The helicopters are another very interesting story, and we know where the helicopters were—they traditionally put their helicopters near some of their other outfits, and we tracked them very carefully. What happened is despite the fact that the Iraqis claim that we indiscriminately bombed civilian targets, they took their helicopters and dispersed them all over the place in civilian residential areas just as fast as they possibly could. Quite a few of them were damaged on airfields, those that we could take on airfields, but the rest of them were dispersed.

Q: You mentioned about the Saudi armed forces. Could you elaborate about their role on the first day?

A: The Saudi Army, as I said, the first thing they did we had this Marine attack that was going through here, and of course we were concerned about the forces over here again, hitting the flanks. That's one of the things you just don't want to have happen to your advancing forces. So this force over here, the eastern task force, had to attack up the coast to pin the enemy in this location. The Saudi forces in this area attacked through here, again, to pin all the forces in this area because we didn't want those forces moving in this direction, and we didn't want those forces moving in that direction. It's a tough mission because these people were being required to fight the kind of fight that the Iraqis wanted them to fight. It's a very, very tough mission. I would point out, it wasn't only the Saudis, it was the Saudis, the Kuwaitis, the Egyptians, the Syrians, the Emiris from United Arab Emirates, the Bahrains, the Qataris, and the Omanis, and I apologize if I've left anybody out, but it was a great coalition of people, all of whom did a fine job.

Q: Is there anything left of the Scud or chemical capability?

A: I don't know, but we're sure going to find out if there's anything left. The Scuds that were being fired against Saudi Arabia came from right here. So obviously, one of the things we're going to check on when we finally get to that location is what's left.

Q: Could you tell us in terms of the air war, of how effective you think it was in speeding up the ground campaign? Obviously, it's gone much faster than you ever expected. As a second part of that, how effective do you think the air/land battle campaign has been?

A: The air war, obviously, was very, very effective. You just can't predict about things like that. You can make your best estimates at the outset as to how quickly you will accomplish certain objectives, but of course, a lot of that depends on the enemy and how resilient the enemy is, how tough they are, how well dug in they are. In the earlier phases we made great progress in the air war. In the latter stages we didn't make a lot of progress because frankly, the enemy had burrowed down into the ground as a result of the air war. That, of course, made the air war a little bit tougher, but when you dig your tanks in and bury them, they're no longer tanks. They're now pill boxes. That, then, makes a difference in the ground campaign. When you don't run them for a long time they have seal problems, they have a lot of maintenance problems and that type of thing. So the air campaign was very, very successful and contributed a great deal.

How effective was the air/ground campaign? I think it was pretty effective myself. I don't know what you all think.

Q: Can you tell us what you think as you look down the road would be a reasonable size for the Iraqi Army, and can you tell us roughly what the size is now if the war were to stop this evening?

A: With regard to the size right now, at one time Saddam Hussein was claiming that he had a seven million man army. If he's got a seven million man army, they've still got a pretty big army out there. How effective that army is, is an entirely different question.

With regard to the size of the army he should have,

I don't think that's my job to decide that. I think there are an awful lot of people that live in this part of the world, and I would hope that is a decision that's arrived at mutually by all the people in this part of the world to contribute to peace and stability in this part of the world. I think that's the best answer I can give.

Q: You said the gate was closed. Have you got ground forces blocking the roads to Basra?
A: No.

Q: Is there any way they can get out that way?
A: No. (Laughter) That's why the gate's closed.

Q: Is there a military or political explanation as to why the Iraqis did not use chemical weapons?
A: We had a lot of questions about why the Iraqis didn't use chemical weapons, and I don't know the answer. I just thank God that they didn't.

Q: Is it possible they didn't use them because they didn't have time to react?
A: You want me to speculate, I'll be delighted to speculate. Nobody can ever pin you down when you speculate. Number one, we destroyed their artillery. We went after their artillery big time. They had major desertions in their artillery, and that's how they would have delivered their chemical weapons. Either that or by air, and we all know what happened to the air. So we went after their artillery big time. I think we were probably highly, highly effective in going after their artillery.
There are other people who are speculating that the

reason why they didn't use chemical weapons is because they were afraid if they used chemical weapons there would be nuclear retaliation.

There are other people that speculate that they didn't use their chemical weapons because their chemical weapons degraded, and because of the damage that we did to their chemical production facilities, they were unable to upgrade the chemicals within their weapons as a result of that degradation. That was one of the reasons, among others, that we went after their chemical production facilities early on in the strategic campaign.

I'll never know the answer to that question, but as I say, thank God they didn't.

Q: Are you still bombing in northern Iraq? If you are, what's the purpose of it now?

A: Yes.

Q: What's being achieved now?

A: Military purposes that we . . . Exactly the same things we were trying to achieve before. The war is not over, and you've got to remember, people are still dying out there. Those people that are dying are my troops, and I'm going to continue to protect those troops in every way I possibly can until the war is over.

Q: How soon after you've finally beaten the Republican Guard and the other forces that threaten you, will you move your forces out of Iraq—either into Kuwait or back into Saudi?

A: That's not my decision to make.

Q: Are you going to try and bring to justice the people responsible for the atrocities in Kuwait City? And also, could you comment on the friendly fire incident in which nine British were killed?

A: I'm sorry, that was asked earlier and I failed to do that. First of all, on the first question, we have as much information as possible on those people that were committing the atrocities, and of course, we're going through a screening process. Whenever we find those people that did, in fact, commit those atrocities, we try and separate them out. We treat them no differently than any other prisoner of war, but the ultimate disposition of those people, of course, might be quite different than the way we would treat any other prisoner of war.

With regard to the unfortunate incident yesterday, the only report we have is that two A-10 aircraft came in and they attacked two scout cars, British armored cars, and that's what caused the casualties. There were nine KIA. We deeply regret that. There's no excuse for it, I'm not going to apologize for it. I am going to say that our experience has been that based upon the extremely complicated number of different maneuvers that were being accomplished out here, according to the extreme diversity of the number of forces that were out here, according to the extreme differences in the languages of the forces out here, and the weather conditions and everything else, I feel that we were quite lucky that we did not have more of this type of incident. I would also tell you that because we had a few earlier that you know about, that we went to extraordinary lengths to try and prevent that type of thing from happening, It's a terrible tragedy, and I'm sorry that it happened.

Q: [Inaudible]

A: I don't know, I'm sorry. I don't believe so because I believe the information I have that a forward air controller was involved in directing that, and that would indicate that it was probably during the afternoon. But it was when there was very, very close combat going on out there in that area.

Q: The United Nations General Assembly was talking about peace. As a military man, you look at your challenge, and you can get some satisfaction out of having achieved it. Is there some fear on your part that there will be a cease-fire that will keep you from fulfilling the assignment that you have? Is your assignment as a military man separate from the political goals of the . . .

A: Do I fear a cease-fire?

Q: Do you fear that you will not be able to accomplish your end, that there will be some political pressure brought on the campaign?

A: I think I've made it very clear to everybody that I'd just as soon the war had never started, and I'd just as soon never have lost a single life out there. That was not our choice. We've accomplished our mission, and when the decision-makers come to the decision that there should be a cease-fire, nobody will be happier than me.

Q: We were told today that an A-10 returning from a mission discovered and destroyed 16 Scuds. Is that a fact, and where were they located?

A: Most of those Scuds were located in western Iraq. We went into this with some intelligence estimates that

I think I have since come to believe were either grossly inaccurate, or our pilots are lying through their teeth, and I choose to think the former rather than the latter, particularly since many of the pilots have backed up what they've been saying by film and that sort of thing. But we went in with a very, very low number of these mobile erector launchers that we thought the enemy had. However, at one point we had a report that they may have had ten times as many. I would tell you, though, that last night the pilots had a very, very successful afternoon and night as far as the mobile erector launchers. Most of them in western Iraq were reportedly used against Israel.

Q: You've said many times in the past that you do not like body counts. You've also told us tonight that enemy casualties were very, very large. I'm wondering with the coalition forces already burying the dead on the battlefield, will there ever be any sort of accounting or head counts made or anything like that?

A: I don't think there's ever been, ever in history of warfare, been a successful count of the dead. One of the reasons for . . . That's because it's necessary to lay those people to rest, for a lot of reasons, and that happens. So I would say that no, there will never be an exact count. Probably in the days to come you're probably going to hear many, many stories—either over-inflated or under-inflated, depending upon who you hear them from. The people who will know best, unfortunately, are the families that won't see their loved ones come home.

Q: If the gate is indeed closed, as you said several

times, and the theories about where these Kuwaiti hostages are—perhaps Basra, perhaps Baghdad, where could they be? A quick second question, was the timing for the start of the ground campaign a purely military choice, or was it a military choice with political influence on the final choice of dates?

A: When I say the gate is closed, I don't want to give you the impression that absolutely nothing is escaping. Quite the contrary, what isn't escaping is heavy tanks, what isn't escaping is artillery pieces, what isn't escaping is that sort of thing. That doesn't mean that civilian vehicles aren't escaping, that doesn't mean that innocent civilians aren't escaping, that doesn't mean that unarmed Iraqis aren't escaping—that's not the gate I'm talking about. I'm talking about the gate that is closed on the war machine that is out there.

The timing for the beginning of the ground campaign, we made a military analysis of when that ground campaign should be conducted. I gave my recommendation to the Secretary of Defense and General Colin Powell, they passed that recommendation on to the President, and the President acted upon that recommendation. Why, do you think we did it at the wrong time? (Laughter)

Q: I'm wondering if your recommendation and analysis was accepted without change.

A: I'm very thankful for the fact that the President of the United States has allowed the United States military and the coalition military to fight this war exactly as it should have been fought, and the President in every case has taken our guidance and our recommendations to heart, and has acted superbly as the Commander in Chief of the United States.

Thank you very much.

APPENDIX B

Service Resume

RESUME OF SERVICE CAREER
OF
H. NORMAN SCHWARZKOPF, General

DATE AND PLACE OF BIRTH 22 August 1934, Trenton, New Jersey

YEARS OF ACTIVE COMMISSIONED SERVICE
Over 32

PRESENT ASSIGNMENT Commander in Chief, United States Central Command, MacDill Air Force Base, Florida 33608, since November 1988

MILITARY SCHOOLS ATTENDED
The Infantry School, Basic and Advanced Courses
United States Army Command and General Staff College
United States Army War College

EDUCATIONAL DEGREES
United States Military—BS Degree—No Major
University of Southern California—MS Degree—Mechanical Engineering

FOREIGN LANGUAGE(S) French, German

MAJOR DUTY ASSIGNMENTS

FROM	TO	*ASSIGNMENT*
Oct 56	Mar 57	Student, Infantry Officer Basic Course and Airborne School, United States Army Infantry School, Fort Benning, Georgia
Mar 57	May 59	Platoon Leader and later Executive Officer, Company E, and later Assistant S-3 (Air), 2d Airborne Battle Group, 187th Infantry, Fort Campbell, Kentucky
Jul 59	Jul 60	Platoon Leader, Company D, later Liaison Officer, later Reconnaissance Platoon Leader and later Liaison Officer, Headquarters and Headquarters Company, 2d Battle Group, 6th Infantry, United States Army Europe
Jul 60	Jul 61	Aide-de-Camp to the Commanding General, Berlin Command, United States Army Europe
Sep 61	May 62	Student, Infantry Officer Advanced Course, United States Army Infantry School, Fort Benning, Georgia
Jun 62	Jun 64	Student, University of Southern California, Los Angeles, California
Jun 64	Jun 65	Instructor, Department of Mechanics, United States Military Academy, West Point, New York

Jun 65	Apr 66	Airborne Task Force Adviser, Airborne Brigade, United States Military Assistance Command, Vietnam
Apr 66	Jun 66	Senior Staff Adviser/G-5 (Civil Affairs) Adviser, Airborne Division, United States Military Assistance Command, Vietnam
Jun 66	Jun 68	Associate Professor, Department of Mechanics, United States Military Academy, West Point, New York
Aug 68	Jun 69	Student, United States Army Command and General Staff College, Fort Leavenworth, Kansas
Jun 69	Dec 69	Executive Officer to the Chief of Staff, Headquarters, United States Army Vietnam
Dec 69	Jul 70	Commander, 1st Battalion, 6th Infantry, 198th Infantry Brigade, 23d Infantry Division (Americal), United States Army Vietnam
Jul 70	Jun 72	Chief, Professional Development Section, Infantry Branch, Officer Personnel Directorate, Office of Personnel Operations, Washington, DC
Aug 72	Jun 73	Student, United States Army War College, Carlisle Barracks, Pennsylvania

Jun 73	Oct 74	Military Assistant, Office of the Assistant Secretary of the Army (Financial Management), Washington, DC
Oct 74	Oct 76	Deputy Commander, 172d Infantry Brigade, Fort Richardson, Alaska
Oct 76	Jul 78	Commander, 1st Brigade, 9th Infantry Division, Fort Lewis, Washington
Jul 78	Aug 80	Deputy Director for Plans, United States Pacific Command, Camp H.M. Smith, Hawaii
Aug 80	Aug 82	Assistant Division Commander, 8th Infantry Division (Mechanized), United States Army Europe
Aug 82	Jun 83	Director, Military Personnel Management; Office of the Deputy Chief of Staff for Personnel, United States Army, Washington, DC
Jun 83	Jun 85	Commanding General, 24th Infantry Division (Mechanized) and Fort Stewart, Fort Stewart, Georgia
Jul 85	Jun 86	Assistant Deputy Chief of Staff for Operations and Plans, United States Army, Washington, DC
Jun 86	Aug 87	Commanding General, I Corps, Fort Lewis, Washington
Aug 87	Nov 88	Deputy Chief of Staff for Operations and Plans/Army Senior Member, Military Staff Committee, United Nations, Washington, DC

PROMOTIONS	*DATES OF APPOINTMENT*	
	Temporary	*Permanent*
2LT		1 Jun 56
1LT	29 Nov 57	1 Jun 59
CPT	24 Jul 61	1 Jun 63
MAJ	28 Jul 65	1 Jun 70
LTC	12 Aug 68	1 Jun 77
COL	1 Nov 75	1 Jun 80
BG	1 Aug 78	22 Jan 82
MG		1 Jul 82
LTG	1 Jul 86	
GEN	23 Nov 88	

US DECORATIONS AND BADGES
Distinguished Service Medal
Silver Star (with 2 Oak Leaf Clusters)
Defense Superior Service Medal
Legion of Merit
Distinguished Flying Cross
Bronze Star Medal with V Device (with 2 Oak Leaf Clusters)
Purple Heart (with Oak Leaf Cluster)
Meritorious Service Medal (with Oak Leaf Cluster)
Air Medals
Army Commendation Medal with V Device (with 3 Oak Leaf Clusters)
Combat Infantryman Badge
Master Parachutist Badge
Army Staff Identification Badge
Secretary of Defense Identification Badge

SOURCE OF COMMISSION USMA

SUMMARY OF JOINT EXPERIENCE

Assignment	*Dates*	*Grade*
Airborne Adviser, Airborne Brigade, later Senior Staff Adviser, G-5 (Civil Affairs), Airborne Division, United States Military Assistance Command, Vietnam	Jun 65–Jun 66	Major
Deputy Director, J-5 (Plans), United States Pacific Command, Hawaii	Jul 78–Aug 80	Brigadier General
Deputy Director, Operation URGENT FURY, Grenada Invasion Operation	Oct 83	Major General

Commander in Chief, United States Central Command, MacDill Air Force Base, Florida	Nov 88–Present	General

As of 23 November 1988